a *Natural* WAY OF BUSINESS

KATHLEEN & UDAYAN GUPTA

Foreword by PRESIDENT WILLIAM JEFFERSON CLINTON

GRUPO PUNTACANA: AN UNUSUAL
PARTNERSHIP IN SUSTAINABLE TOURISM

a Natural WAY OF BUSINESS

HOW FRANK RAINIERI, THEODORE KHEEL, OSCAR DE LA RENTA & JULIO IGLESIAS HELPED TRANSFORM AN ISLAND ECONOMY

KATHLEEN & UDAYAN GUPTA

Foreword by PRESIDENT WILLIAM JEFFERSON CLINTON

GONDOLIER

A NATURAL WAY OF BUSINESS : GRUPO PUNTA CANA AND THE DEVELOPMENT OF SUSTAINABLE TOURISM
© 2006 Udayan Gupta and Kathleen Zaborowsky
Photography © 2006 David Lane
Published by: Gondolier, a division of Bayeux Arts, Inc., 119 Stratton Crescent SW,
Calgary, Canada T3H 1T7
www.bayeux.com

Book design by David Lane
Copy edited by Tim Carroll

Library and Archives Canada Cataloguing in Publication

    A natural way of business : Grupo Punta Cana and the development
of sustainable tourism / editors, Udayan Gupta and Kathleen Zaborowski ;
photography by David Lane.

ISBN 1-896209-99-8

    1. Punta Cana Resort and Club. 2. Grupo Punta Cana. 3. Ecotourism--
Dominican Republic—Case studies. 4. Sustainable development—Dominican
Republic—Case studies. I. Gupta, Udayan, 1950- II. Zaborowski, Kathleen, 1950-

HC153.5.Z7P86 2004   338.4'79172930455
C2004-905861-4

First Printing: June 2006
Printed in Canada by Friesens

All rights reserved. No part of this publication may be reproduced, stored in a retrieval system, or transmitted, in any form or by any means, electronic, mechanical, recording, or otherwise, without the prior written permission of the publisher, except in the case of a reviewer, who may quote brief passages in a review to print in a magazine or newspaper, or broadcast on radio or television. In the case of photocopying or other reprographic copying, users must obtain a license from the Canadian Copyright Licensing Agency.

The Publisher gratefully acknowledges the financial support of the Canada Council for the Arts, the Alberta Foundation for the Arts, and the Government of Canada through The Book Publishing Industry Development Program.

Books published by Bayeux Arts, under its Bayeux and Gondolier imprints, are available at special quantity discounts to use as premiums and sales promotions, or for use in corporate or organizational training programs. For more information, please write to Special Sales, Bayeux Arts, Inc., 119 Stratton Crescent SW, Calgary, Canada T3H 1T7.

Metropolitan College of NY
Library - 7th Floor
60 West Street
New York, NY 10006

PHOTOGRAPHY BY DAVID LANE

WITH KATHLEEN & UDAYAN GUPTA

AND JAKE KHEEL

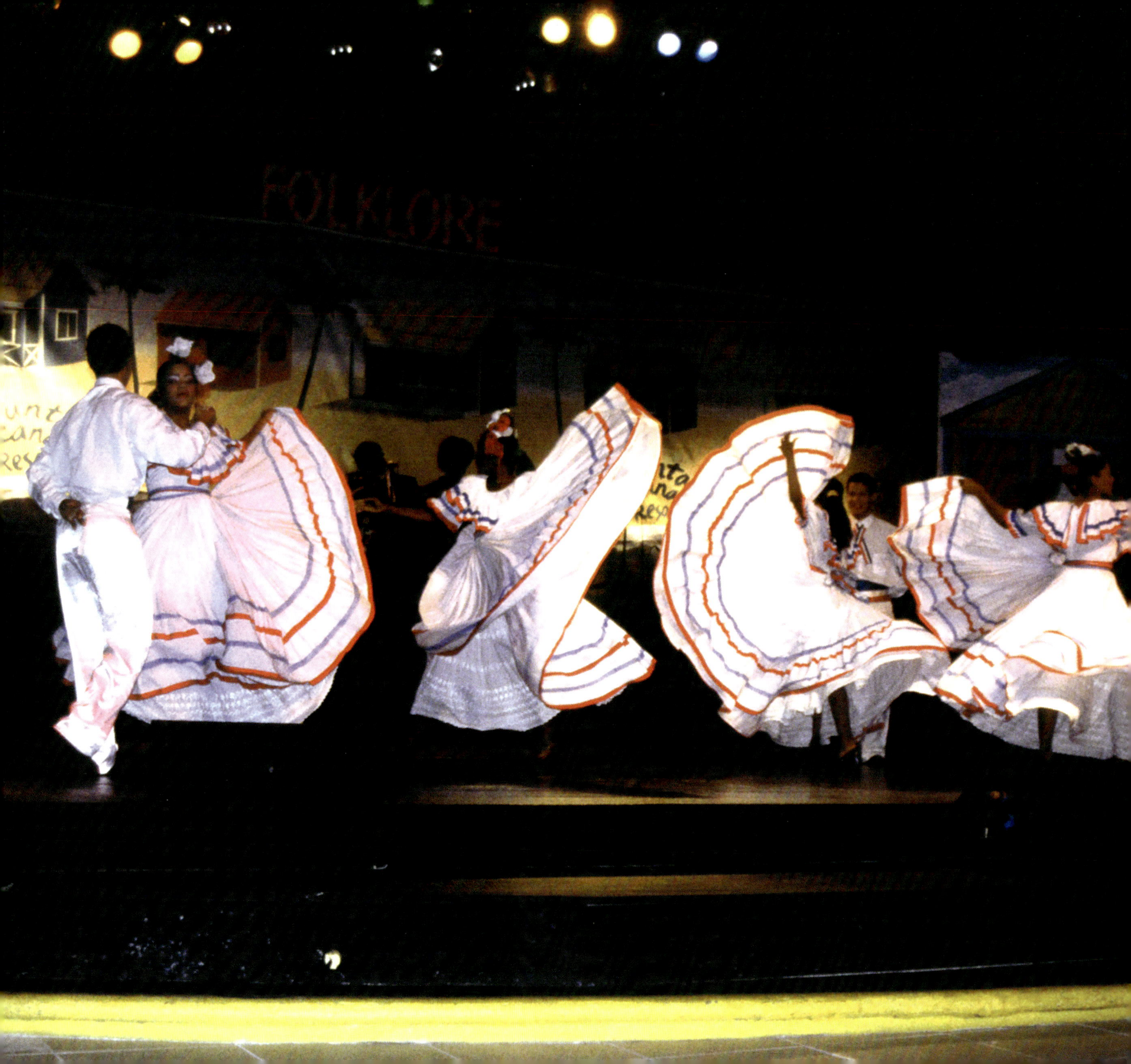
FOLKLORE

Contents

Foreword
By President William Jefferson Clinton **ix**

Introduction **xi**

CHAPTER 1
Grupo Punta Cana: An Unique International Partnership in Sustainable Tourism **1**

CHAPTER II
Two Partners, Two Perspectives **15**

CHAPTER III
Vision into Plan/Plan becomes Reality **29**

CHAPTER IV
Conserving Nature/Building Human Capital **45**

CHAPTER V
Human and Economic Development; Workers and Entrepreneurs Building Together **67**

CHAPTER VI
Moving Forward/Blending Vision and Reality **87**

Epilogue **99**

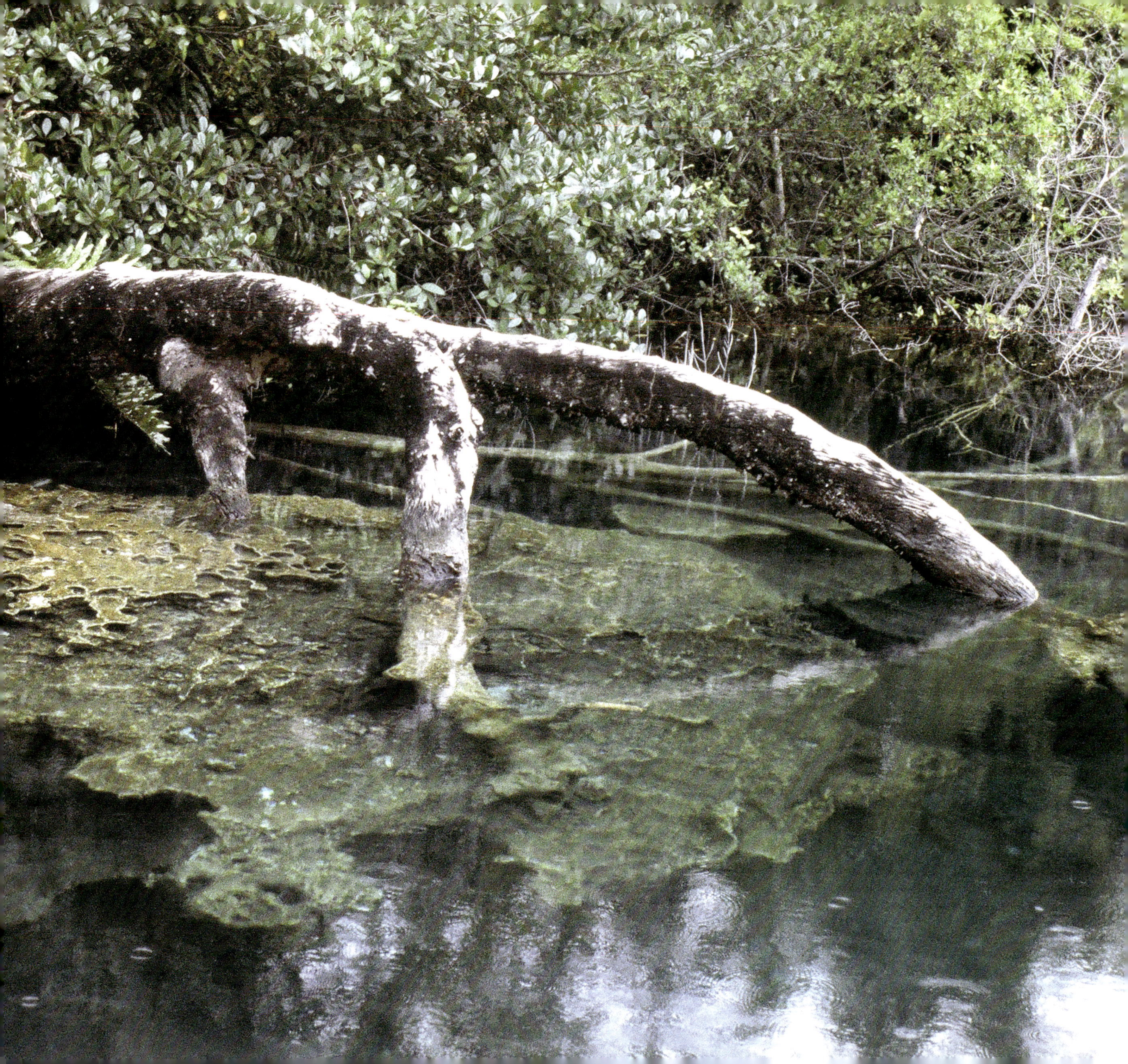

Foreword

President William Jefferson Clinton

I FIRST VISITED THE DOMINICAN REPUBLIC'S PUNTA CANA Resort and Club as a guest of my good friend Oscar de la Renta. It's an exquisite place, with beaches of fine, white sand, palms swaying in the wind, and the clear blue ocean extending to the horizon.

With its natural beauty and friendly, welcoming people, Punta Cana is special. But what Frank Rainieri, Ted Kheel, Julio Iglesias, and Oscar de la Renta have done there makes the resort even more extraordinary—they've found a way to share Punta Cana with the world while preserving it. The Resort and Club was designed to delight guests while bringing infrastructural improvements to the residents of Punta Cana. Their approach, which has become known as sustainable tourism, protects the environment and the community, employs local workers, and still manages to ensure a profitable enterprise.

In many tourist destinations, a natural tension exists between development, which draws more visitors but often destroys what they come to see, and environmental protection, which protects the appeal of the destination while reducing profit potential. The Punta Cana Resort and Club strikes the perfect balance between the two.

The first small resort was built at Punta Cana in 1971, and by the following year Frank and Ted established a school for the resort workers' children. They continued to expand over the years, but they didn't turn their back on the community. In 1994, they established the Punta Cana Ecological Foundation. This non-profit organization manages 2,500 acres, which include an interpretive center and a network of nature trails.

Since then, the Ecological Foundation has opened a biodiversity lab that collaborates with American universities and Dominican community organizations.

Frank and Ted, plus their newer partners, Julio and Oscar, continue to take special care of their employees. In recent years, they have built housing for resort staff, a church, and an elementary school. This year, they completed the first phase of a high school for up to 500 students from the surrounding community. This is what sustainable tourism is all about—by tending to the beauty of their land and the people who live there, they maintain its viability as a tourist destination and at the same time do the right thing for Punta Cana and its residents.

I encourage you to read this book and learn for yourself how it all happened. It's a good story, but it's also an important example of what can be achieved by people who are concerned with helping others as well as making a profit. I hope Punta Cana will become a model for other developers, and a destination for more tourists. A visit there is made far richer by the knowledge that you are supporting such a worthy endeavor.

—President William Jefferson Clinton

Intoduction

IN THE SUMMER OF 1972, *The Wall Street Journal* carried a front-page article by Jonathan Quitney poking fun at a fledgling resort on the east coast of the Dominican Republic. The thirty square miles of then undeveloped land had been purchased by a group that included several notable Americans involved in labor and management, more as a lark than a serious investment. But Quitney wrote that they were running a cotton plantation and paying substandard wages, when, in fact, there was only one man trying unsuccessfully to grow cotton.

The failed cotton plantation was all the way back in 1970—before the tiny seed of 10 cottages were built on the land, and more importantly before Ted Kheel, the leader of the group that had purchased the land, turned for help to Frank Rainieri, a young Dominican who had just returned from schooling in the United States. Over the next three decades, Rainieri and Kheel set about to create a model enterprise that not only proved Quitney wrong, but also has become a paradigm for sustainable development.

In 1994, some 22 years after Quitney's jibe, as I was writing for *The Wall Street Journal*, I met Ted Kheel, who quickly said, "You're just the man I wanted to meet. Years ago your newspaper made fun of our investment in tourism. Now we are a financial success. Don't you think that *The Wall Street Journal*, of all publications, should take note of our capitalistic achievement?"

Kheel was right. Under the headline "Jeers Turn to Cheers," *The Wall Street Journal* carried my account of the success that Punta Cana has become.

evolving—is a remarkable study in sustainable development.

From the very beginning, Rainieri and Kheel chose the more difficult path. Build the resort and the community together—in lock step. You want your employees to be living within the community, not at a commuting distance that is an hour away. You want their children to grow up within the community and benefit from the resources that have been created. The value of a vibrant community isn't only in the wealth and businesses created, but also in the intangibles—work-force loyalty, easier recruitment, and a proud and self-reliant people.

In 1994, Rainieri and Kheel added a new wrinkle to the entire enterprise when they established an ecological foundation with 1,500 acres of land. Ostensibly, the foundation was designed to draw attention to the environment, an issue vital to the Dominican Republic. But with researchers coming to the foundation from major universities in the U.S., the foundation and its facilities have become an important destination for environmental scholarship. At Punta Cana, there is now the unique opportunity to observe research and implementation side by side. For many Dominicans this is also an opportunity to study their own environment and their own ecology alongside some of the world's leading experts.

On a summer evening in Punta Cana you can attend an informal presentation by people as well-known as Harvard professors E.O. Wilson and Brian Farrell addressing the mystery of the Caribbean ant plagues that devastated local agriculture and ecology centuries ago. Or you can wander the 2,000 acre ecological park or scuba dive and snorkel the beachfront and

In the decade that followed, Punta Cana has evolved into a dynamic community—a community that has not only developed on its own, but also anchored the development of the entire region. As development experts and economists debate and puzzle over strategies to create businesses, create jobs and revitalize economies in less-developed countries, the Punta Cana model must be seen as among the very attractive.

In taking a community in which there was no sanitation or drinking water, where literacy levels were at frightening lows, and which had no means of quickly changing those education levels, a region in which a central government lacks the means to provide support, Rainieri and Kheel were taking extraordinary risks. Perhaps there were risks the two men weren't even aware of. But the outcome—which is still

CONSERVING NATURE/BUILDING HUMAN CAPITAL

the eight kilometer coral reef that parallels Punta Cana.

Currently on the drawing board is a cultural center, part of a cultural interchange that will promote Dominican culture both in Punta Cana and in the U.S. It is part of Punta Cana's efforts to organize its own cultural venue, away from the traditional dominance of Santo Domingo.

In a world where economic development is more rhetoric than practice and models of development are almost always cast in the image of Silicon Valley and Massachusetts' Route 128, Punta Cana stands out as a model of success that can easily be emulated.

More important, it stands out as a paradigm of sustainable development in which the primary actors aren't the only beneficiaries. The beneficiaries include an entire country—a region that is emerging from the throes of abject poverty and chronic underdevelopment, a regional economy that has seen the revitalization of existing businesses and the creation of new ones, local communities that now have running water, sanitation and electricity and access to educational and health opportunities they never had before.

Ted Kheel likes to talk about the process of economic development in Punta Cana as a process of interchange—not a superimposition of one process on another or just an amalgam of ideas—that brings together the strengths of the Dominican Republic and the United States. It is not a cookie-cutter model, but one that has been derived from trial and error. It truly is how America can best help its neighbors.

A Natural Way of Business is the story of Punta Cana and sustainable development told by the dozens that have been part of the experience. Some of the storytellers—Frank Rainieri, Ted Kheel, Oscar de la Renta, and Julio Iglesias—are formal stakeholders. Others are family members who expect to continue to develop the vision that began three decades ago. And then there are the dozens that in their own ways have been part of this grand experience. Together, their narratives provide a unique insight into a new way of development.

CONSERVING NATURE/BUILDING HUMAN CAPITAL

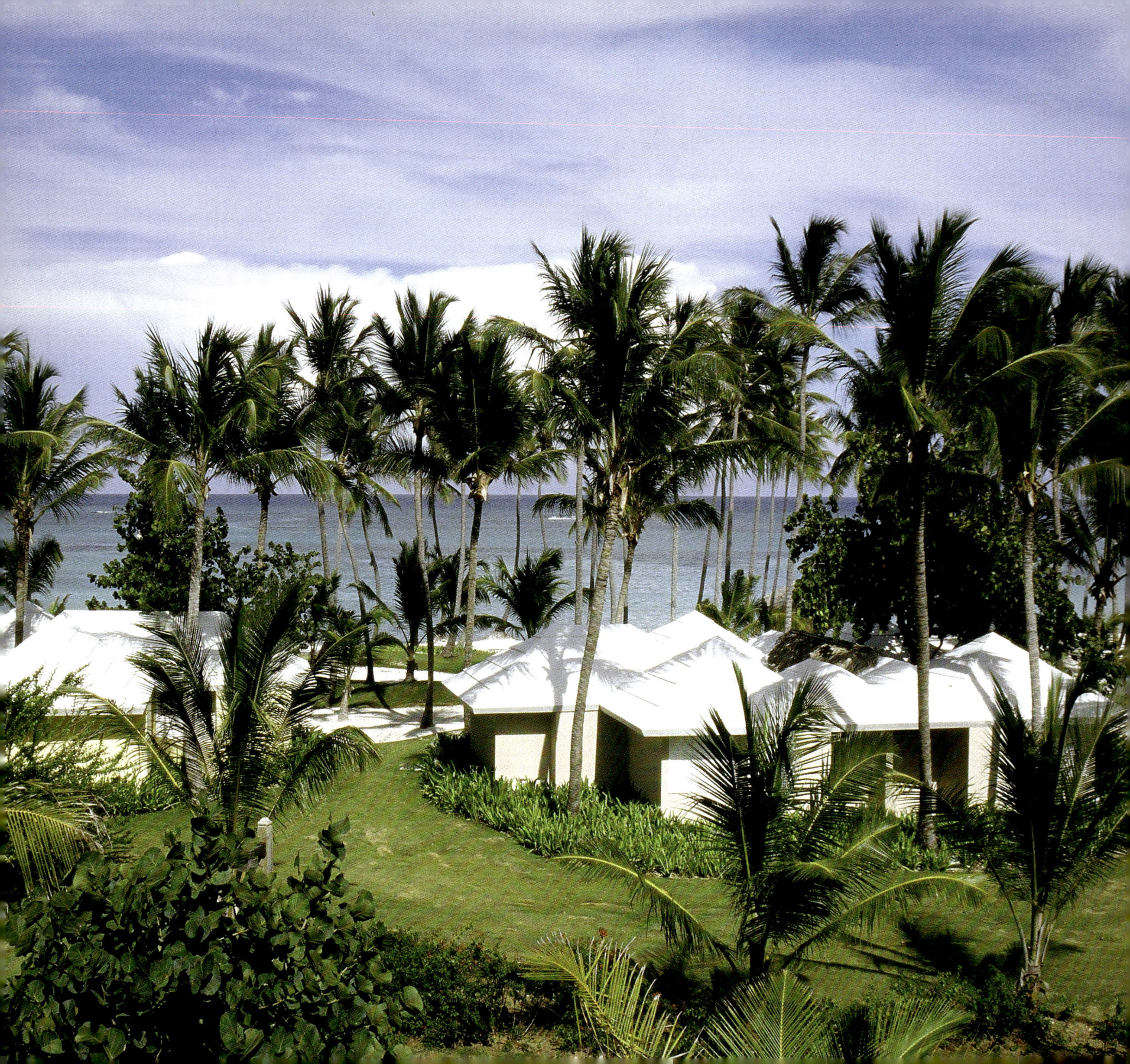

Grupo Punta Cana: A Unique International Partnership in Sustainable Tourism

No one ever heard of the place!
The following are brief excerpts from newsletters that Ann S. Kheel sent to members of her family.

November 1969: *This weekend Ted went from Florida to the Dominican Islands with Keith Terpe of Puerto Rico to look at land.*

November 1971: *Family news is plentiful, Dad and I managed a happy weekend visit to Punta Cana. . . . The country's President Balaguer flew in and Dad greeted him . . . Because we will be there Xmas, I would like each of you to write down the following information for it has no phone . . . No one ever heard of the place . . .*

THIRTY FIVE YEARS LATER, Punta Cana is no secret any longer. It is one of the Caribbean's most prized resorts, an elegant community that not only hosts million of tourists a year, but also is home to celebrities such as Oscar de la Renta, Julio Iglesias, and Mikhail Baryshnikov, Once a poor fishing community whose inhabitants burned wood to sell as charcoal, the Punta Cana area now boasts its own airport—the third busiest in the Caribbean, a commercial plaza, designer boutiques, schools, and a unique biodiversity laboratory that attracts researchers and students from some of the most prestigious universities in the U.S. In addition, the resort itself features a unique 7,152 yard golf course, one of the longest in the Caribbean, that sits opposite a sandy beach dotted with tall coconut palms fronting the Mona Passage.

Build a community and you'll have built a business. That's Punta Cana.

In thirty-five years, a restaurant owner and a labor lawyer have partnered to help transform what was at best a marshy wasteland in a remote corner of the Dominican Republic into a major development cornerstone.

What Frank Rainieri and Ted Kheel have achieved in Punta Cana is immense. Not only have they succeeded in putting together a business that is both successful and profitable in spite of mishaps, struggles, and setbacks such as hurricanes

Dancers perform in traditional Domincan costumes. The show continues with modern versions of several Dominican themed dances and concludes with a spectacualr fire show.

Grupo Punta Cana:
A Unique International
Partnership in
Sustainable Tourism

and bank failures, they have taken a community that was once raw jungle and transformed it into one of the Caribbean's most popular destinations. In the process, they created thousands of new jobs, new business entities, and brought in much-needed hard currency and partners.

By bringing Oscar de la Renta and Julio Iglesias into the ownership of Grupo Punta Cana, Rainieri and Kheel have created a unique international partnership. Neither Iglesias nor de la Renta is a veteran of the resort business, but each man brings a visibility and an international perspective to the resort that money simply can't buy.

Oscar de la Renta provides the aesthetic. A Dominican-born designer who is known and admired internationally, he has chosen not only to build his home in Punta Cana's Corales, but also to invest in the entire project. In one fell swoop he brought a new image to an area that was once described as "some real estate along the island's scrubby and largely featureless eastern coast," with no major towns and an array of "all-inclusive" hotels catering mainly to down-market jumbo-jet crowds from Düsseldorf.

Julio Iglesias, the Spanish-born singer, who began his working life as goalkeeper for the famed Real Madrid and switched careers after a car accident, has created a mystique and romance around Punta Cana that has since attracted others to the area, including Mikhail Baryshnikov and President and Mrs. Clinton.

The two men together provide a new international perspective. De la Renta and Iglesias bring their eye and their experience to Punta Cana. Because they travel far and wide, including to other resorts and hotels, their experience has become a great tool in helping the resort at Punta Cana understand what it is doing right and what it can do better.

Julio Iglesias

WHEN I FIRST SAW CORALES it was almost jungle. When we first got there on the bumpy roads, Miranda, who was pregnant with our first child, asked if this is what it was going to be like all the time. But the whole place is so intriguing and so beautiful that you can't help but fall in love with Punta Cana. The people are amazing. They are a generous group and have a quality of the countryside: friendly and open.

The love affair with Punta Cana begins when people

come in. The airport is very beautiful; the land surrounding the airport is very beautiful. This is a perfect place for a vacation, for second homes. What I believe most is that in this part of the world there is a great deal of attraction in nature, the beaches, the sea, the land. But we have to help keep the environment whole. We have to help keep the reef preserved. A lot of the reef is under danger, pressure from construction. But when the oil tankers go by they leave a lot of damage in their trail, especially on the reefs.

One of our biggest concerns is the environment. We need to respect the ecology of the place and keep the land in shape. Keeping the natural things in shape is important. We are always watching and taking care of the trees, the plants, and the use of water.

I have always believed that the way we can contribute to Punta Cana and the Dominican Republic is by taking care of the people who take care of the land. We need to make sure that they have access to education and they grow with us because of the opportunities that are there. It's about giving dignity to people and in a way that all of us can benefit.

The success of Punta Cana will be when the people who live there have good housing, when their kids are all getting good education—the ones that keep on staying there and moving there because of the increased opportunities. Punta Cana isn't just about the beaches, the golf courses, and the resort, it is about creating a community and really learning how to build a community that is lasting where there is a future for everyone.

For too long Punta Cana was just there, with no growth and no development. The Rainieris have done a remarkable job in changing its profile. The main thing for us in Punta Cana

The restaurant at the Marina.

THE LOVES OF JULIO IGLESIAS

PORT OF CALL, Dominican Republic, Aug. 14, 2004

Julio Iglesias doesn't really mind going home, and who can blame him when home is a six-acre compound in the resort town of Punta Cana in the Caribbean Island nation of the Dominican Republic.

Iglesias humorously admits to doing nothing there.

"Sometimes to do nothing, for me, is to have two hours of floating in the water, which is quite a paradise, and thinking, 'Oh God, what a lucky man I am,'" says Iglesias.

GRUPO PUNTA CANA: A UNIQUE INTERNATIONAL PARTNERSHIP IN SUSTAINABLE TOURISM

is to be attractive to other people—not only in the rest of the Dominican Republic but to the rest of the world. We have to play the role of developing the country. Fortunately, the country is a lot better today economically and business-wise. In Punta Cana alone we provide jobs to as many as 5,000 people and the plans call for adding 10,000 more.

The Rainieris and the Kheels have been most instrumental in coming up with a plan and executing it in a manner that few people can. And now we are going ahead expanding the community, filling in the areas, helping provide sanitation, water, and housing. This is not a matter of devotion; this is an obligation. This has to be done if we are to grow and develop.

The Rainieris are there, Oscar is very much involved with the education. What I feel I can do is tell people how to keep the land the environment in shape.

Oscar de la Renta

In the Columbia Electronic Encyclopedia there is a simple citation:

de la Renta, Oscar , 1932–, *French fashion designer, b. Santo Domingo, Dominican Republic. He studied in Madrid and began a career in fashion with* <u>Balenciaga</u>*. Moving to France in the 1960s, he became known for his luxurious clothes, especially evening wear, of extravagant materials. He designed for Elizabeth Arden in New York and also established his own firm that produced romantic and opulent clothes. His designs encompass everything from bathing suits to wedding dresses, furs, perfumes, and linens.*

Although over the last three decades or so, Oscar de la Renta has operated mostly out of the U.S., he also returned to the country he left when he was 18 years old. He built his first home at Casa de Campo, but in 1994, together with singer Julio Iglesias, bought into Grupo Punta Cana and decided to build his house in Corales. He now is deeply immersed in the planning and development of Punta Cana, as well as in the marketing.

WHAT WE ARE TRYING IS to control the areas that we own, not only to protect the foreigners or the Dominicans who live in the resort. But what is most important is to protect the people who come in to work here. They are very simple people. I don't think there is any other project in the DR that creates schools and educates the children of the employees. That doesn't exist anywhere else. In the DR we are quite an unusual type of operation. Like Frank and Haydee, I am Dominican and I care for my people. Punta Cana is my future home and I am hoping, eventually, that I can spend a lot of time here. I want to create a healthy environment for my family and all of the people around me. The people who work for me are my extended family. Fernando Soler, for example, is like my son. He is a wonderful success because he confirms to me that if you help people they can make it.

You have to create an environment for people to succeed. One of the most difficult things about Punta Cana was that there was no life for a family or an infrastructure for family life. One of the first things with Frank and Ted—when we got involved—was not to make the place just for visitors, but to create a successful environment for the workers so that they will be happy to live here and live here with their families.

I moved from Casa de Campo to Punta Cana for several reasons. My house in Casa De Campo was the very first house built there. I was sort of a pioneer. I loved the project. I loved the idea. I have lived away from the Dominican Republic for 50 years, but to me where I come from continues to be very

A private house on the golf course.

important. I built that house in Casa de Campo because I wanted a place that I could enjoy with my friends and I lived there for 28 years. A couple of things happened: I felt that Casa de Campo was getting a little too congested. I wasn't able to control my environment in Casa de Campo. I felt that there was too much construction. I am a visual person and I hated a lot of the houses that were being built there. It is the most successful resort in the Caribbean from the point of view of the people that come there, compared with other places such as Cancun and St. Bart's. Casa de Campo attracts a higher echelon of international society, but I felt that the place was becoming too crowded.

Julio Iglesias is a very old friend of mine, very much like my brother. And he was always telling me, "Why don't you find me some place to build a house?" There was no land left in Casa de Campo. So finally I came to the decision that it was time for me to move on. I called Julio and said to him: "This is the time for you and me to have a house next to each other." We flew down to the Dominican Republic. I knew I wanted to be in the Dominican Republic because of the beaches and the sun. It's the kind of life that I like. I thought I had never met Frank Rainieri. I didn't know, but I had met him many, many years ago when he came to my house in Casa de Campo. I didn't know about his particular project. That was in the late '70s, early '80s when Punta Cana was just beginning.

I had heard what was happening in Punta Cana, that there was an untapped area with beautiful beaches and still relatively unspoiled. So I called Frank Rainieri, introduced

GRUPO PUNTA CANA: A UNIQUE INTERNATIONAL PARTNERSHIP IN SUSTAINABLE TOURISM

myself and said I was going to come down to Punta Cana with my friend Julio Iglesias. We are looking at the possibility of making an investment and building ourselves homes. My call was in 1994. I knew all of the beautiful beaches were on the east coast. So we flew down to the Dominican Republic to the airport in Punta Cana. We had the helicopter and we went all over the eastern coast. What was most attractive about Punta Cana, especially to Julio who is lucky enough to own his own plane, was the fact that Punta Cana was five minutes away from an airport. After our helicopter ride, we started to talk to Frank. And the idea began to develop. At first he showed the area closest to the hotel. That was too close to the hotel. Then we went to the other area, Corales, but we couldn't go very far because there was no access. When I started to finally build my house there we had to open up the path with machetes because there was no other access. To get water to the construction site we had to get the fire trucks from the airport to provide water to mix the cement.

Originally, Julio and I were going to be involved only in the area of Corales and we were going to be partners with Ted and Frank in the development of Corales. Then the opportunity came. The airport was then owned 50% by Ted and Frank and 50% by Club Med. But Club Med wasn't doing too well financially. And one of my very closest friends, Giovanni Agnelli, was chairman of Fiat and the largest shareholder of Club Med. So I talked to Frank and told him that perhaps we could buy out the Club Med share. So Ted and I flew to Paris and negotiated to buy the 50% of the airport so we would be in total control of the airport. The airport has been a tremendous success. It services all of the tourists in the area and it has allowed us to grow at a faster pace. Though Punta Cana has become a business for me, I didn't originally go there for business.

Can we build a larger economic entity out of Punta Cana? I think so. The most difficult growth times are over.

Casa de Campo was successful. It was part of a big town and it had the human resources to feed its growth. We didn't have that luxury. The closest town to us is Higuey, which is about 40 to 45 minutes by bus. So we needed to create a life for the people to come and work in Punta Cana, a place where not only the employees felt comfortable, but also where they felt they could bring their family. A family-safe environment.

I haven't lived in the DR, but I have had the privilege of living in other countries and visiting a lot of the world. So in a lot of ways I have lived a privileged sort of life. I know a lot of

GRUPO PUNTA CANA:
A UNIQUE INTERNATIONAL
PARTNERSHIP IN
SUSTAINABLE TOURISM

people and I have been to a lot of places, so I know some of the things I would like for Punta Cana, and also some of the things I do not want to happen. We can create an example of how to develop tourism and something that is balanced for everyone.

I would love to see Punta Cana as an example of what you can do to develop a business that serves three purposes. It creates a safe and enjoyable environment for those who visit it. Second, and most important, it protects that environment. Third is the human aspect of it all.

A hotel owner I recently met visited Punta Cana for the first time in 2004. As someone in the industry, he had been aware of the area for some time. But it was the presence of Oscar de la Renta and Julio Iglesias that encouraged him to visit, saying, "I wanted to see how they helped set Punta Cana apart."

Indeed, there are many who now see Punta Cana as an extremely viable resort community, not simply to visit but also to stay. Kheel and Rainieri started to build a business and found they needed to create a community. Now that the community has been created, they find they have a sustainable business.

It's not been easy. Tourism is an extremely complex and deplete-able asset. In most tourism communities around the world, tourists are transient customers who show little or no interest in the community around the asset. In fact, in most Third World environments, there often is a vast schism between the tourist and the immediate community.

On the African coastline as well as in the Caribbean, the tourism business often involves three distinct constituencies—tourists, the local resident community, and resort and tour operations run by multinational companies with headquarters in the U.S. or in Europe. While the resort and tour operators are interested in quality labor, they aren't interested in actively helping to develop the community from which that labor is extracted. Indeed, a developed community directly around a tour area runs counter to the financial interests of developers and operators: It often increases the cost of labor and the cost of the product being offered to the tourist.

That is why the efforts of Rainieri, a Dominican, and Kheel, an American lawyer deeply involved in championing labor, are so significant. They could have chosen to develop a business without developing a community. But they would also have to deal with the boom-and-bust spending characteristics of tourism. In season, tourist spending buoys the economy. Out of season, the economy languishes. If you can't

The Punta Cana bowling alley.

create year-round spending, you can't sustain the business and the community.

Rainieri clearly understood that developing the community was not counter to the interests of the resort business. In fact, enhancing the buying power of the community—through year-round residents and through a more affluent local community—could be just the answer. It wasn't simply a matter of giving the locals a few loaves and few fishes in the hope that they would multiply, it was a matter of teaching them how to bake bread and how to fish.

✹

Cornell Professor David M. Stipanuk, in writing about Punta Cana, says:

> Despite being an attractive area for tourism development—white sand beaches, groves of coconut trees, coral reefs, and vegetation—no supporting infrastructure was in place. There was little electricity, no waste-treatment facilities, not enough potable water to support a growing population, and no local education facilities. The area had no nearby airport and access highways didn't exist. The region had only a marginal economic base consisting of some commercial fishing, subsistence agriculture, and charcoal production, which can lead to long-term environmental devastation of a community. (It takes 30 large trees to produce one 80-pound sack of charcoal.) Illiteracy rates ran as high as 70 percent, and job opportunities were limited.
>
> To see Punta Cana as a future tourism destination required great vision and determination. It also required a commitment to the notion of sustainable tourism. Access roads and an airport had to be built, a supply chain for

food and other amenities was needed, schools had to be built, and access to clean water and a reliable power supply had to be developed.

Rainieri was not concerned that creating educational and training opportunities for local residents could actually take them away—to other cities or even to the U.S. for better prospects—because his master plan called for the creation of similar "advanced" jobs within Punta Cana. And he counted on the desire and the need of many of these people to stay and grow within the region because things were better. To satisfy the demand for a better trained work force—a demand that would grow over the next decade or so—you didn't need to import higher-priced transient workers. You simply had to train your own. More important, if you created the infrastructure for your work force to prosper—running water, sanitation, housing, schools, health care and schools—they would stay and become the core of a local community.

Adds Stipanuk:

> Sustainability—development that enables local residents to achieve higher standards of living and preserves communities, cultural and environmental assets—is a continuing journey, not a destination. The Punta Cana Resort and Club, located on the east coast of the Dominican Republic, demonstrates what can be accomplished when private business, the government, educational institutions, and the local community collaborate on such a journey.

While sustainability continues to be a work in progress (as should be true for all tourism operations), the Punta Cana Resort and Club has greatly contributed to the revitalization and relative improvement of the economic health of an area once considered marginal.

It has accomplished this through a variety of business initiatives, including adapting technologies, training personnel, and creating environmental and education programs.

While sustainability is good for business and the community the business is located in, it doesn't necessarily play well with the customers—the tourists. They want the white sand beaches, the unspoiled coral reefs, the coconut groves, but they also want an escape, a fantasy. The idea that they are in the same community in which Oscar de la Renta, Julio Iglesias, and Mikhail Baryshnikov live, and the Clintons visit, creates that fantasy and that escape.

Iglesias's involvement has grown with time. The locals talk about his plane and his family and his rock-star son with admiration and respect, adding to the mystique. And the singer himself has made himself more available to help champion and promote the Grupo Punta Cana's expanding initiatives.

De la Renta has been even more pro-active. The 72-year-old designer talks openly about moving to Corales not simply for vacationing but for retirement. And he has begun to take an active role in working with local entrepreneurs in helping hone their talents and their businesses. His involvement with the Oscar de la Renta boutique at the airport and with the Bana store at the Plaza and his mentoring of their owners, Eladia Torres and Fernando Soler, is an important step in creating Punta Cana's entrepreneurial culture. Indeed, in recruiting de la Renta and Iglesias as stakeholders in Grupo Punta Cana, Rainieri and Kheel have gone out of the box, but also have created a unique partnership.

The final brilliance of the Punta Cana model is its ability to establish itself as an ecological/biodiversity destination. After a few days at a resort—after a few days of the sun and the sand and the food and the wine, tourists get bored, says

Grupo Punta Cana: A Unique International Partnership in Sustainable Tourism

Cornell's Eloy Rodriguez. "And what better way to stimulate them and challenge them than by involving them in the environment they are consuming?" he asks.

Writes Stipanuk about protecting the environment:

> Since the early stages of its development, Punta Cana Resort and Club incorporated environmental concerns and it continues to demonstrate keen sensitivity to these issues. The resort's facilities are set back from the beach and existing trees were preserved in the construction phase. Development is low density, boasting 525 square meters of area per guest at full occupancy.
>
> The overall design and architecture of the resort, created by Dominican architect Oscar Imbert, incorporates Dominican, Spanish, and Awark Indian themes, using local materials when appropriate. On a regional basis, guidelines have been established for clearing land and maintaining green areas, and all residential as well as commercial developments are required to comply with these environmental standards.
>
> At the recently developed Punta Cana Golf Club, the use of seashore paspalum hybrid grass permits the environmentally sound practice of combining seawater and recycled fresh water to irrigate the grounds. Recycled water also is used to irrigate the hotel's gardens.
>
> Located immediately adjacent to the resort is the 2,000-acre Punta Cana Nature Reserve. The Punta Cana Ecological Foundation was established by the resort to oversee this reserve as well as to fund the Punta Cana/Cornell Biodiversity Center.
>
> In collaboration with the local technical university INTEC-Instituto National Technologico de Santo Domingo and the Cornell College of Agriculture and Life Sciences, the foundation launched a program to identify and inventory the region's flora and fauna. It also conducts environmental education programs for both staff and visitors, including activities to celebrate international environmental days, tree-planting campaigns, beach cleanup operations, and bird observation outings. And a guided tour of the reserve is now offered to hotel visitors.
>
> The foundation's alliance with Cornell, culminating in the creation of the Cornell Biodiversity Laboratory in Punta Cana, is a notable achievement. The Biodiversity Center houses a 5,000 square foot laboratory, a teaching facility, and a dormitory used by students and faculty from Cornell, Harvard, Columbia, as well as Dominican universities. It conducts inventories of plants, animals, and marine and microbial organisms that may lead to new medicines one day. There is also ethno-medicine research, which involves collecting information from citizens in rural villages, who for generations have used natural compounds as "home remedies," and studying those compounds to determine their makeup.
>
> In 1995, GPC and the United Nations co-sponsored with Earth Pledge Foundation and Earthkind International the first Caribbean Conference on Sustainable Tourism. GPC has continued to be active in hosting similar events, including a conference on Caribbean biodiversity (July 2001), travelers' philanthropy (November 2002), and the conference "Making Biodiversity Work for Your Travel Business" (April 2003).

As the business grows and the resort develops, involvement in the community becomes key to moving forward. Says Antonio Ramis, general manager of the resort: "We want everyone coming here to have the same will to do things. If we cannot, it is not worth it. We want them to care about social issues such as education health and housing. We are opening the high school building next. We will start preparing the people to create a new village for our people to move into. Punta Cana is moving in that direction."

Development is a double-edged sword: In the process of creating jobs and creating wealth, you can easily "corrupt" a culture. How do you grow an economy and preserve its culture?

Developing tourism is especially tricky. Tourism is about selling the climate, the culture and the environment to a transient customer, a customer who has little at stake in that indigenous culture and environment. Developers, in order to maximize their returns from tourists, will often run down the very resources that are the key elements of successful tourism. Successful tourism requires that the conflict between development and conservation be resolved creatively. And Punta Cana is one of the unique examples of how the conflict has been resolved. Punta Cana demonstrates that a successful resort can also be the cornerstone of a dynamic indigenous community.

It all began randomly. A piece of land that couldn't be used productively. A vision, rather than a plan. The airport brought it all together. Then began the building of a resort

Grupo Punta Cana:

A Unique International

Partnership in

Sustainable Tourism

community with the recognition that it's not enough to create a transient population. You have to create a year-round community. And that involves more investment, more than a resort, building for the locals.

Conservation has been the major glue. Not only in shaping the business, but also in creating a sense of awareness, a center for learning, and a connecting point for the area. The biodiversity laboratory has not only become a place for learning to conserve, but it also has become a central research facility in the area.

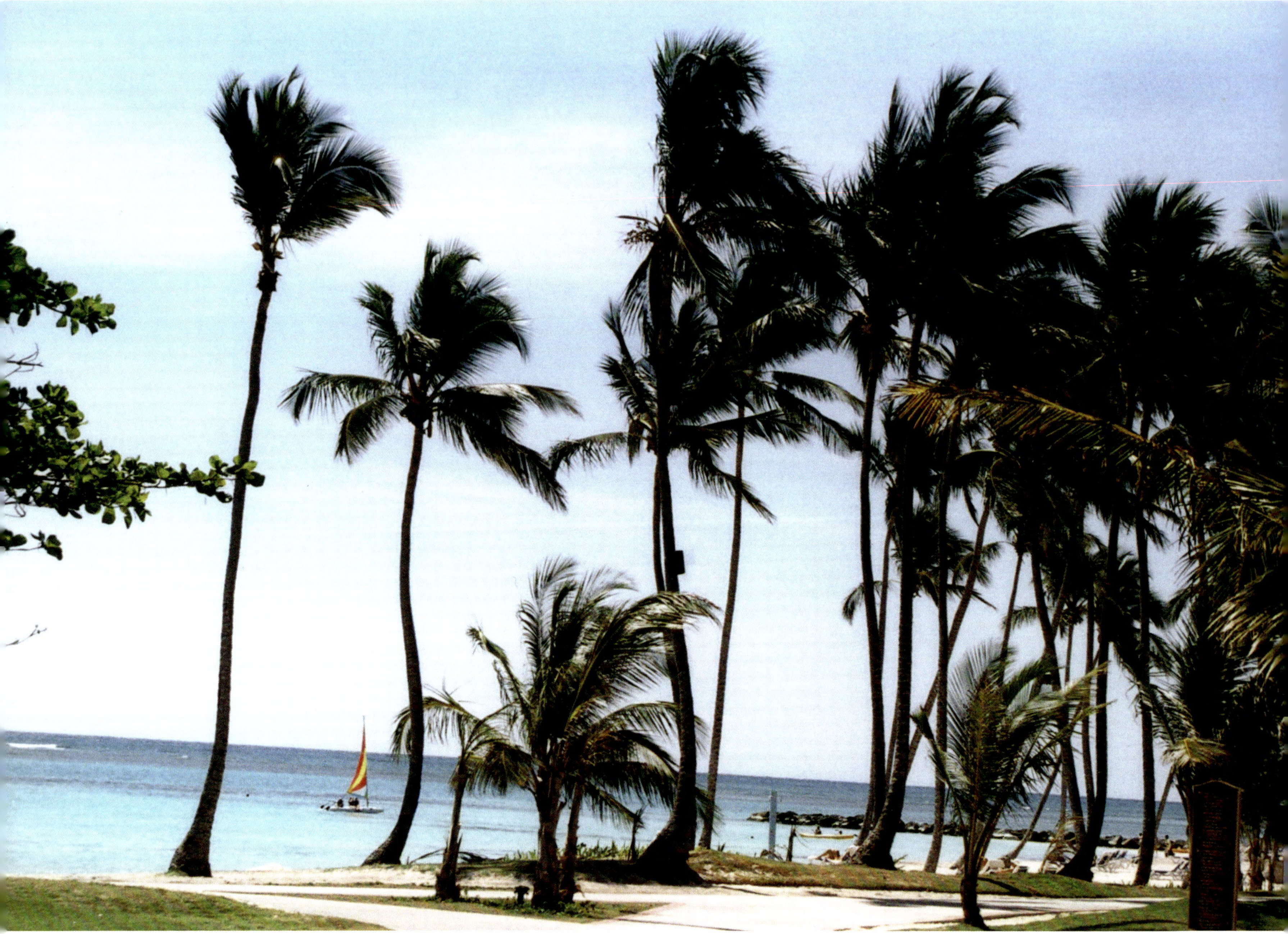

Two Partners, Two Perspectives

Business primers repeatedly extol the strength of partnerships and how much work needs to go in to find partners who can work together and grow. The task is even more complex when the partners are from different cultures, different backgrounds, and have roots in different business experiences. The partnership between Frank Rainieri and Ted Kheel is proof of how critical and synergistic partners are in creating and building a business, especially one as time-consuming and complex as Punta Cana.

In searching for partners, many often look for those who are mirror images of themselves. After all, compatibility is a key element in being able to work together, experts say. What may be more important to a partnership is "complementarity." What are the partners' strengths? How do they complement each other? What do they bring to the table as individuals and as partners?

The manner in which Rainieri and Kheel have worked together is a classic in learning about partnerships.

Theodore Kheel

I GOT INVOLVED WITH THE DOMINICAN REPUBLIC by accident. Thirty-five years ago, I was attending the midwinter meeting of the AFL-CIO, which was held in Miami, a meeting that I attended regularly. It was a conference of the unions but it was also attended by people who did business with the unions. Banks would be there. Pension funds would be there. Companies that provided medical care would be in attendance. Mediators, arbitrators would come to these meetings.

I was there and one of the union people said he had been to the Dominican Republic where land was pretty cheap. This was the 1960s, the aftermath of the assassination of Trujillo—a man very much like Saddam Hussein, a dictator who ran the country, killed people he disagreed with. But within the Dominican Republic there was significant opposition to Trujillo. He was aware of it and he was constantly guarding himself against the effort by his enemies to do anything about it. But those who were his enemies concluded he had to go. And they conspired with the CIA and were given advice and ammunition on how to assassinate Trujillo. One night they finally succeeded.

The conspirators were counting on a general in the Dominican army to lead the army in revolt, but he changed his mind at the last minute. The conspirators shot Trujillo, put his body in the back of a car, and went looking for this general. In the meantime, one of the people in the shootout went to the hospital because he had been wounded in the arm. When they asked him what happened to him, he couldn't explain it. So they called the secret police. The secret police tortured him, and he confessed to the plot to kill Trujillo and that they had succeeded. Trujillo's sons returned to the Dominican Republic from Europe and the country was in a state of great uncertainty, very similar to the conditions in Iraq at the present time. The circumstances were different in that after a time the country settled down.

In any event, this was the period when we learned that land was very cheap and why don't we buy a couple of acres? We could then say that we owned a few acres in the Dominican Republic. This fellow said he'd look around and see what was available. He returned and said a parcel of 30 square miles was available for $200,000. We put together a group of people and bought the land. And then we found that it wasn't worth what we paid for it.

The area was raw jungle, no infrastructure, no water, no roads. We didn't know quite what to do with it. We met up, ironically with the army general—whose name was Imbert, the last surviving member of the cabal that assassinated Trujillo. Most of the group had been tortured and killed except for Antonio Imbert, who hid in the apartment by arrangement with Frank Rainieri's father, who was of Italian origin and had friends in the embassy. He stayed there for six months until things settled down. When he came out, he found out he was the only survivor. His son is our architect. What happened was his daughter was engaged to marry Frank Rainieri and she and her mother flew to Puerto Rico to buy a trousseau. The plane crashed and they were killed. Frank had been studying in the U.S. and had just returned to the Dominican Republic.

Antonio Imbert, the general, whose daughter and wife had been killed, said to this union leader who was part of our group, "Do you have anything for our almost son-in-law?" The guy said: "Yes, he can run these 30 square miles for us." So we engaged Frank Rainieri and he turned out to be a superior hotelier, a businessman. What we have now is a tremendous success. It is his accomplishment and my support. He couldn't have done it without me, and I couldn't have done it without him.

This kind of collaboration is the key to business in developing countries. You need the support that comes from people who can bring in dollars. And the advice that comes from knowledge and experience in business matters. But you also need someone locally who is able to be respected for his ability and knowledge of the local situation. Frank was the indispensable person in charge and I was the CEO providing philosophy, providing financial support, providing strategic decisions.

The initial $200,000 financing was for the purchase of the land. We got 50 or 60 people to put up money. Nobody put up more than $5,000 or $10,000. Most people put up $1,000. We formed a Dominican company called the Coddetreisa, which was an acronym for a company engaged in business and we bought the property in the name of that company. That company name since then has been changed to Grupo Punta Cana.

Under the old laws of the Dominican Republic, a foreigner cannot own property without the approval of the president. But a foreigner can own stock in a Dominican company. We sold units in Coddetreisa to about fifty, sixty people, and used that money to build a small resort. It consisted of ten bungalows and a lodge. Each bungalow had two bedrooms and two beds in each room. So if you multiplied four people in each bungalow by ten bungalows, the capacity was forty people. We never had forty people. The most we'd ever have was during Christmas time when I came with twenty members of my family and we still had room for more.

Punta Cana was a delightful place. It was like a private estate. And people who came there said, isn't this lovely. Why don't you keep it this way? Well, it was totally uneconomic and it was then that Jonathan Quitny of *The Wall Street Journal* sneaked in and wrote a front-page article that said that George Meany, Lane Kirkland, and Ted Kheel were engaged in running a cotton plantation. Quitny visited when Frank was experimenting with growing cotton. One man with his wife and a couple of kids were planting cotton to try to make it grow. That became a cotton plantation in *The Wall Street Journal*.

We had engaged a man named Charlie Cahill, a Puerto Rican, to run the property. Then one day, we got a phone call from an American who said, "We just bought your property but you can buy it back for $50,000." I said, "As a matter of principle we are not disposed to be shaken down." But, in addition, we didn't have the $50,000. We retained Frank's relative, Jose Machado, who was a professor of law, to represent us. The first thing that this man shaking us down discovered was that under Dominican law he was not allowed to own the property. So he transferred the property to a Dominican named Troncoso. So we started a lawsuit against Troncoso and against Charlie Cahill, who left the country and never returned because if he returned he would probably be arrested for illegally transferring the property. We had made him the

TWO PARTNERS, TWO PERSPECTIVES

17

president of the company and he purported to have sold all of the property to this Trancoso—first the American, then to Trancoso. So we went to court in El Seibo.

They say if you go through El Seibo faster than fifteen miles an hour, you'll miss it. And we were in litigation there for a year. Then the decision came down saying that the property belonged to us. They took that to the land court of three people and we won the decision there. They took it to the Supreme Court and we won there also. That turned out to be terrific. We had been told that we would have no chance in the Supreme Court. I went to the Secretary of State William Rogers, who was a classmate of mine, and said I don't expect you to send our troops to the Dominican Republic to recapture our property, but I want to be sure you have a record of what is happening. He put me in touch with the Dominican desk of the State Department. So I told the story to the Dominican desk and we won unanimously in every court. When we got the final decision I wrote a letter—by that time Henry Kissinger had become secretary of state—we were Americans in the Dominican Republic and we got a fair shake.

Now that they see this is a most successful resort, they say you are a visionary. I have to say we are the beneficiaries of the mistakes that many people made. This potentially disastrous lawsuit earned us the respect of the Dominican community. First of all, we didn't get shaken down. Secondly, we won unanimously in three courts.

It was after our victory that Club Med came along and agreed to do a joint venture with us. They loved bargains and they paid us $300,000 and a percentage of the shares in the new company we formed with Club Med. They built their resort and then found it inaccessible. You could get there but it was such a chore. They went to the government and asked for an airport. And the government said we'd be delighted to build one but we don't have the money. So we went to them and said: "Why don't we build an airport?" We went to the Overseas Private investment Corp (OPIC), a U.S. corporation that lends money to Americans to invest in developing countries. They said they could not lend to us because we had never built an airport and they didn't know if we will be able to finish it, if we could be operational, and if it could work properly. So we came back to Club Med and created a joint venture where we put up the land and they put up the money and we guaranteed that with $1.5 million we would build a 5.000 foot runway with a terminal building and have it in operation in eight months. In December 1993, we opened the Punta Cana International Airport.

But even that airport was too small to be satisfactory for Club Med's purposes because passengers had to fly to San Juan and then change to a smaller plane. They could not fly overseas directly. We returned to OPIC and got a loan for $1.6 million from OPIC that enabled us to extend the airport to 7,200 feet. Then something happened that we totally didn't expect: We discovered that our market was not the U.S., but Europe. Planes began to fly to the Dominican Republic from Europe. They could cross the ocean and land at 7,200 feet, but with only 7,200 feet of runway, they could not take off fully loaded. So they would have to make an intermediate stop at Puerto Plata.

Europe. Punta Cana. Puerto Plata. Punta Cana. Europe. That wasn't satisfactory, although it worked. We then had some cash flow and we extended the airport runway to 9,200 feet, and then to 10,400 feet.

The airport is the most unusual aspect of what we have done. We benefited from Club Med's mistakes in building a resort without an airport and then funding it. We are the third-largest airport in the Caribbean. And we are far and away the largest airport in the Dominican Republic. We bring in more money into Santo Domingo than any other airport and there are now 40 hotels in the area.

After the airport, we set out to build a resort. That was done with a group of Dominicans who joined forces and we set aside 105 acres and built a resort that is much larger. We did it through some creative financing under a law to encourage construction. The Dominican government passed a law that gave us tax benefits. If the owners of condos agreed to make their units available to tourism and would not occupy it themselves for more than ten weeks a year—the other forty-two weeks would be made available to tourists—they would get significant tax breaks.

We built seventy-nine villas and five buildings with studios. Each building had thirty-two studios. We built villas that had two or three bedrooms. And we sold them. The owners agreed that they would occupy their building not more than ten weeks a year and would put the building into the rental

TWO PARTNERS,
TWO PERSPECTIVES

pool. We would furnish the units so that they were all the same so that the visiting public would not know that a certain villa belonged to Ted Kheel. They were all furnished the same way. I bought quite a few at the time. We had overexpanded. I bought a building and eight villas. And I still own a building of thirty-two units and have eight villas. And that was indispensable financing when it was done. Also, the unit owners got 10% of net rentals. In this country you buy condos and furnish it yourself. Here they bought the condos and we furnished it. And the owners would get 10% net rentals on their investment. We could have financed it conventionally by selling the units and people would furnish it themselves. But instead the owners got ten weeks of stay and 10% of rentals.

We did creative financing with a number of companies in order to expand. The currency laws then said—they've changed since—that you could take out of the Dominican Republic in dollars only 18% of your investment a year. So if you were Colgate-Palmolive or Texaco or ARCO, you came there and wanted to conduct business, you had to bring in your dollars and set up an office. You had to do the business in pesos and the government was maintaining a one-to-one ratio so that if you brought in a million dollars, the Central Bank would take the million dollars and give you a million pesos. So you conducted business with the pesos. If you made money, you could repatriate up to 18% of your investment. So if you invested one million dollars you could repatriate $180,000 by taking the pesos to the Central Bank, which would then give you dollars. But many of the companies made more than 18% and they would have blocked pesos. They could not take the excess profits out of the country. So they agreed to sell the pesos to Club Med and Club Med agreed to pay them out of the country. So Club Med bought seven million dollars worth of pesos that had the same value as dollars by paying these three companies seven million in the U.S. and they would

Two Partners,
Two Perspectives

then transfer the pesos to them. They got the pesos and paid for it in dollars without interest. It was a good deal all around. All this was approved by the Central Bank of the Dominican Republic.

AIG, the insurance company, insured a company that was selling machinery in the Dominican Republic. They agreed that the company would be paid in dollars. And the Central Bank refused to pay. They had to pay the insured company five or six million dollars and they went to the central bank for the dollars. They finally gave them a deal that they would give the insurance company the dollars if AIG reinvested 50% of the discount in the Dominican Republic. We worked out that deal with the Central Bank. The insurance company invested that money with us and they bought six lots from us. And so we got the money that way. That was a method of paying off the debt by reinvesting in the Dominican Republic. We did a number of other things of that sort to finance what we accomplished.

Frank Rainieri

THE TRUE STORY IS, what we began in 1969 has become the most important tourist destination in the whole Caribbean, second only to Cancun in the whole eastern border of Mexico, the whole of Central America, and the whole Caribbean.

All that is within 30 minutes from our airport, and everything was built after the airport. It took us six years to get the permits and, really, they gave us the permits because they never thought we were going to do it—build the first international, private airport in the world. The president told me, "How much is the government going to have to put up?" and I said, "Nothing." "What's the commitment of the government?" I said: "None." "And what's the government going to get?" "The same thing that you get in any other airport," I said. And he said, "OK. I'll give you the permit." But he never thought we were going to do it.

Within those 20 miles today we have 22,000 rooms for rent and we generate—within those 20 miles—25% of the foreign exchange in the Dominican Republic. This area has the highest per capita income in all of DR, and zero unemployment. Zero. Today more than 40% of the work force comes from other regions of the DR, because this area doesn't have the human resources to provide for the tourist industry.

The second good thing about what were doing is that tourism is using a raw material that doesn't end. You can put a computer factory here but the day that somebody offers a better work force, they just move to another place. Tourism is not just the work force. It is nature and location, so nobody can move the tourism from us. They used to say that tourism was fragile. I remember in the 1970s, people spoke about a fragile tourist business. And I always said, "What is the main industry of Spain, Italy, and Greece?" Three big European countries. More than that, in the process of development, we are now talking about an industry for Third World countries. Tourism gives you the advantages of natural resources that don't end, sun and sea. You just have to be careful not to disturb. And it's going to be there and people are going to be looking at vacation in the Caribbean today, as they were 50 years ago, and will look in another 50 years. So it's a resource that we in Punta Cana have to exploit, and we have visions to be the first ones.

We are an example of what the private sector can do, with practically no help from the government of the DR. Do you know what the total investment of the government of the DR in those 20 miles in the last 34 years? In the year 2004 alone, the government collected $70 million in taxes from us.

Opposite page: Punta Cana airport interior.

Do you know how much the government has invested in 35 years? Five million and that includes roads, water, and electricity. Everything has to be done by the private sector. That's good and that's bad. Good, because it's made the private sector put their money where their mouth is. Bad, because there are areas like education, health, infrastructure, roads, and regional plan that don't exist because the government has never spent the money.

Punta Cana began with a group of Americans, partnered with Ted Kheel, who decided to chip in and buy a piece of property in DR, where 99% of them had never been before. Once the property was bought, they didn't know what to do with it. They imported an American to come to see what could be done over here, and he came with some very bright ideas, about exporting the white sand to PR to make plaster. Of course, no one would allow it. This was jungle—there was nothing in this area. Then they came up with another idea. Let's cut the wood and export the wood. I knew one of the original owners, and he asked me to join and help this guy. I was 23 years old. At that age you say and do anything you'll never do when you're 50.

What they had to do here, I told them, was to first make a road to get there. Second, put some cottages to stay overnight, and third, let's clear a plot so we can land an airplane, a small plane, and that's how I began in Punta Cana. I told Ted Kheel and the old group about what I thought could be done, and that was in New York. I didn't know Ted until that day. They were asking this guy questions and he didn't have answers, and I came up with these ideas, and at the end of the meeting Ted called me over and asked me for my card. A week later, Ted called me, and tells me, " Frank, I want to see you. Can you fly to St. Thomas for the weekend?" He had a house there. I told him I could come for Saturday. So I went and Ted Kheel was with Lane Kirkland, who at that time was secretary/ treasurer of the AFL-CIO. And he asked me, "What would you do if you were in charge?" and I said, "First thing. Buy bulldozers." How much does a bulldozer cost? And I'll never forget at that time it cost $48,600, and a truck to take it to Punta Cana cost another $50,000. And he said, "What do you need a bulldozer for?" I said, "To make a road there. There is no road." He said, "What's the second thing?" "We have to clear an area, make some bungalows so we can stay there overnight, because it's four hours from the closest town." He asked, "What's the possibility of having the government build the roads?" I said forget about it. It's not a priority for them. He asked how much 10 cottages would cost, and I told him. Third, we have to buy two electrical generators. There you have to generate your own power. "Electricity, oh, how much will that cost?" Fourth, let's clear a path so we can land a small plane. "How big is that?" Two thousand feet at least, so we can land a six-passenger, single-engine airplane. And how much will that cost? $250,000. "OK, I agree with that," he said.

"Now, how much do you make?" And that was the only time I lied to Ted Kheel in my life. I was making $275 a month, and I told him $800 a month, and Ted told me, "OK, we want you to come and work for us, and we'll pay you $1,200 a month." I looked at him and said, "Sorry, Ted. I always said I would never work for anybody, unless I was a partner." So he said, "OK, we'll sell you shares." I said, "I don't have money to buy shares." He asked, "Do you have family? Your family will buy." And I said, "I always said, that whatever I would make, I would make it on my own, not with my family money." So Ted, this incredible man said, "I'll pay you $1,000 in cash, and I'll give you $200 in shares every month, and if you make what you said could be made, in the time table, and the amount of money you say, we'll give you another 2.5% of the shares as a bonus." And that is how everything began.

Two Partners,

Two Perspectives

That's a beautiful beginning because it was just starting. I remember the first year we had the money to finish everything. But a few years later, of course, with just twenty rooms in the middle of the jungle you don't make money. And the U.S. economy was changing and not for the good. So the partners start thinking maybe they made a mistake. We opened the cottages in 1971. We started building in August 1970. We opened it formally in October 1971. By 1974, we realized we're not going to make money with that, and the dream that we have a beautiful property, was just a dream, and the partners start saying they wanted to sell their shares. And Ted felt responsible for getting all those people to buy shares. So he started buying back, but we had an operation over here, and I wasn't getting any cash to cover maintenance of this place. So that's how we started negotiating with Club Med to get them involved here and get this place moving.

Club Med was already in Guadalupe. In November 1975, Ted and I went to Paris and started talking to Club Med, and they said, "We'll go there, but you have to get us an airport, a road, water and electricity, and the money to build it." After

Partners (left to right) Frank Rainieri, Haydee Rainieri, Ted Kheel, Julio Iglesias and Oscar de la Renta.

TWO PARTNERS, TWO PERSPECTIVES

three years of negotiation, in May 1978, we were able to sign with Club Med, to build a club in Punta Cana exactly in the same place where we had built the original Punta Cana Club. Club Med decided that was the place that they wanted, which meant we had to close our little place, tear down the ten cottages to get Club Med. Meanwhile, money was short and Ted had to use his money to repurchase his shares from everybody. For seven years I didn't get a salary, because there was no money, but for me the important thing was for me to keep this place alive, because the squatters were coming in, and the only way to protect the property, 1,500 acres, was by having people here and having something going on.

In 1979, we were able to start building a Club Med in Punta Cana, with only one thing in their favor. I was able to convince the government to start the road to Punta Cana, because what I had done was build a patch through the beach, so that we could come here, but it was 15 miles long. The new road was going to be 30 miles along, a real road. About the airport, we told Club Med, "We are going to work with you to get the government to approve it."

We had to use a lot of ingenuity. It was the time before equity swaps, and we came up with the idea that there were a lot of blocked funds in the DR, funds from foreign companies. There was a law that allowed foreign companies to take out of the country only 18% of their returns on their profits. Many had accumulated much more than that. They had it in banks, they didn't reinvest it. So we went to the Dominican government and told them that all that short-term money sitting in banks was only creating a devaluation of the DR peso, because importers were the ones borrowing it to import things, but that there was nothing productive about the money. So we were able to get the president of DR to allow those companies to lend to Club Med on seven years term at 2.5%. Club Med was to repay the loan with the profits Club Med Punta Cana was going to make. At the same time, Club Med made a deal that they would bring into the Dominican economy X amount of dollars per passenger that came to the DR, and they would bring it through the central bank of DR. It was a net profit for DR because instead of having a lot of pesos sitting in the bank generating nothing, it was money that was going to be invested in a hotel that would generate foreign exchange. So we did the first equity swap, in 1978.

It was a creative business package what we were doing, and we got them to allow us to do that. We told Club Med, here we have the money to finance the hotel. So Texaco, ARCO, and Colgate-Palmolive came up with money from their blocked funds to build the Club Med, and Club Med forgot about the airport because they said we got cheap money, cheap land. We got the road and a commitment for the water. Their eyes went wide open, and they went ahead and signed it, and started to build the Club Med. After Club Med started upgrading, we became limited partners of Club Med; we left part of the money they were supposed to pay us invested in Club Med. We were minority partners and didn't have the right to say anything. By 1983, we were desperate, we were out of money, Club Med didn't pay us a dividend, and we wanted to get out. We had debt so we sold our shares to Club Med, and then we paid the banks with that money. Then, with some of that money, we started to make the airport. We trusted a very simple study, because we didn't have money to make it, and I remember I told Ted, I was ready to start the airport but I needed $50,000, and he looked at me and said, "Are you going to make an airport with $50,000?"

I said, "Ted, I don't know, but give it to me and let's see what's happens." I got Oscar Imbert, the architect, who was just out of college, to design us a very small, inexpensive terminal. So Oscar designed the first stage of that airport, and that's why we used a thatch roof: We didn't have money to build a concrete building, or air condition it. After we built the first stage we said: "Wow! Wow! This is the most beautiful thing we have thought up, and that's why we kept on doing it, and still do it the same way today. And, of course, with the same architect.

That is a creative business plan. After I started with $5,000 to clear the way to the runway, Club Med came back to us and said they were willing to make a partnership with us on the airport. The agreement was that they were going to fund half of the airport, we would put up the land and the design, and they would come up with $1.5 million to build the airport. And if we could not deliver the airport in a million and a half, Ted and I would have to come up with whatever amount of money that was necessary. That was the contract we signed. Of course when we opened it, it was a 5,000 foot runway. The first year, we had 2,460 passengers.

In 1987, we went to the Overseas Private Investment Corp. They said we have heard that you have opened your small airport, and now not only is there Club Med, but there are three hotels under construction. They said we are willing to finance you $1.5 million for the extension. The extension cost the same as the original deal—the difference was that we had to use more asphalt for our jets, Boeing 727s and things like that. So we opened in 1987, then suddenly we went from 6,000 to 13000, to 26,000, 65,000, to 125,000, to 165,000, and we kept on going and last year we had 1.3 million arrivals.

After we built the first Punta Cana club, then the airport and the Punta Cana Beach Club—the original name, now everything is the Punta Cana resort—we realized the next thing you have to do when you own 15,000 acres of land is develop real estate. So we began to build the marina. That was the first stage. And we were doing something small, and we had just completed the marina when we had our first hurricane in 1998 that came through this part of the country. Normally, hurricanes come though Barahona or Santo Domingo, never this far east. The hurricane wiped us out just at the moment when we were starting our real-estate experiment. In 1998, our priority became rebuilding the hotel and the airport, and we lagged behind the marina and the real estate, but soon we realized there was tremendous potential in the real estate over here. Why? The DR is only two hours away from Miami, three hours from New York. With good communication, that airport makes a difference. We have people from France, Spain, Italy, all over the U.S. Why? Because they jump in a plane from Paris to Punta Cana, Milan, Punta Cana, Rome to Punta Cana, Vancouver, and every major city and once they get here they are five minutes away from the airport. That opens the whole thing.

We are a beach and golf destination and much more. Normally when tourists go to the Caribbean they go to a beach destination, or a golf destination, but our mix is unique. They can have a house 400 yards from the beach with a golf course in front of their house, and that's not an easy thing.

The big thing that is so important besides security, good investment, communications, etc., people want to retire, and people who buy houses are in their late forties or fifties, close to retirement, and over here you can get help for your house very easily. You can hire people who will cook for you. There are not that many countries in Europe or here where all these conditions come together.

As people are getting very involved in the real-estate part, we're trying to provide them with the club experience, make them understand that to us you're not a Social Security number. I always use that expression because when I went to college in the U.S. my friends at that time were always talking about how they were going to sign with the major U.S. companies. They were going to make $25,000 and I had to come over here and work for $250 a month, and they could not understand that. I told them they were always going to be a Social Security number, the U.S. at that time was a 200 million people society. I said when I come to the Dominican Republic I'm going to be familiar here. I will get to have a name because we are a small country (today we have eight million people, but at that time four million). There I was going to be a Social Security number. Here I'm Frank Rainieri.

I always tell people that story. People want to be treated not as a number but as a person. Make them personalize that experience, call them by their name, and treat them as they are part of our family. If we treat them as such, they will become part of the Punta Cana family. And that's why we always use the term "Punta Cana family." That's why we allow fathers and sons, cousins to work for us. We have here six members of one family working for us. We encourage that. That's why we promote housing, school. That why we promote them to be part of the Punta Cana family, because that's the only way people will understand that our visitors have to become part of our family.

I believe in sustainable development, sustainable tourism, and sustainability. Why? Because we still have a million people that go to bed with one meal a day. You know you have to make things perfect, but you also have to realize that you can't make everything perfect. Some issues you don't fight. You have fishermen and people who know how to do only that thing, and unless you can provide them with a fishing job, and a fishing alternative, they keep on doing it. You can take them to jail three times and it won't work. What I've tried to do in Punta Cana, and what I think is best for the DR, is to realize there is a thin line that separates us and we have to walk it. How much can we preserve? How can we use what we preserve? What I have tried to do is make the two of them work for the benefit of the people. That's why at the same time that we preserve nature and we have a biodiversity center, we have a school. We're building another school for the people, because if we spend all the money preserving nature and we don't put the money to educating people, we're just wasting our time because when people are hungry, or when they are not educated, there is no way that they are going to preserve nature.

Before there was only the economic aspect of things, now there is the social, and a biodiversity aspect that you have to take into consideration when you are making a business decision. So now you have three lines that you have to take very much into consideration.

I have been very fortunate because Ted Kheel and I think similarly. When nobody was talking about sustainability back in the late '80s before the world summit, we were involved in it here. We were starting it. Why? Because we saw the decline of the ambiance in this area. And Ted has always been a vanguard. Ahead of his time. I always thought we both have the same way of thinking, so we have been able to push that way.

Under the canopy at the plant nursery.

TWO PARTNERS, TWO PERSPECTIVES

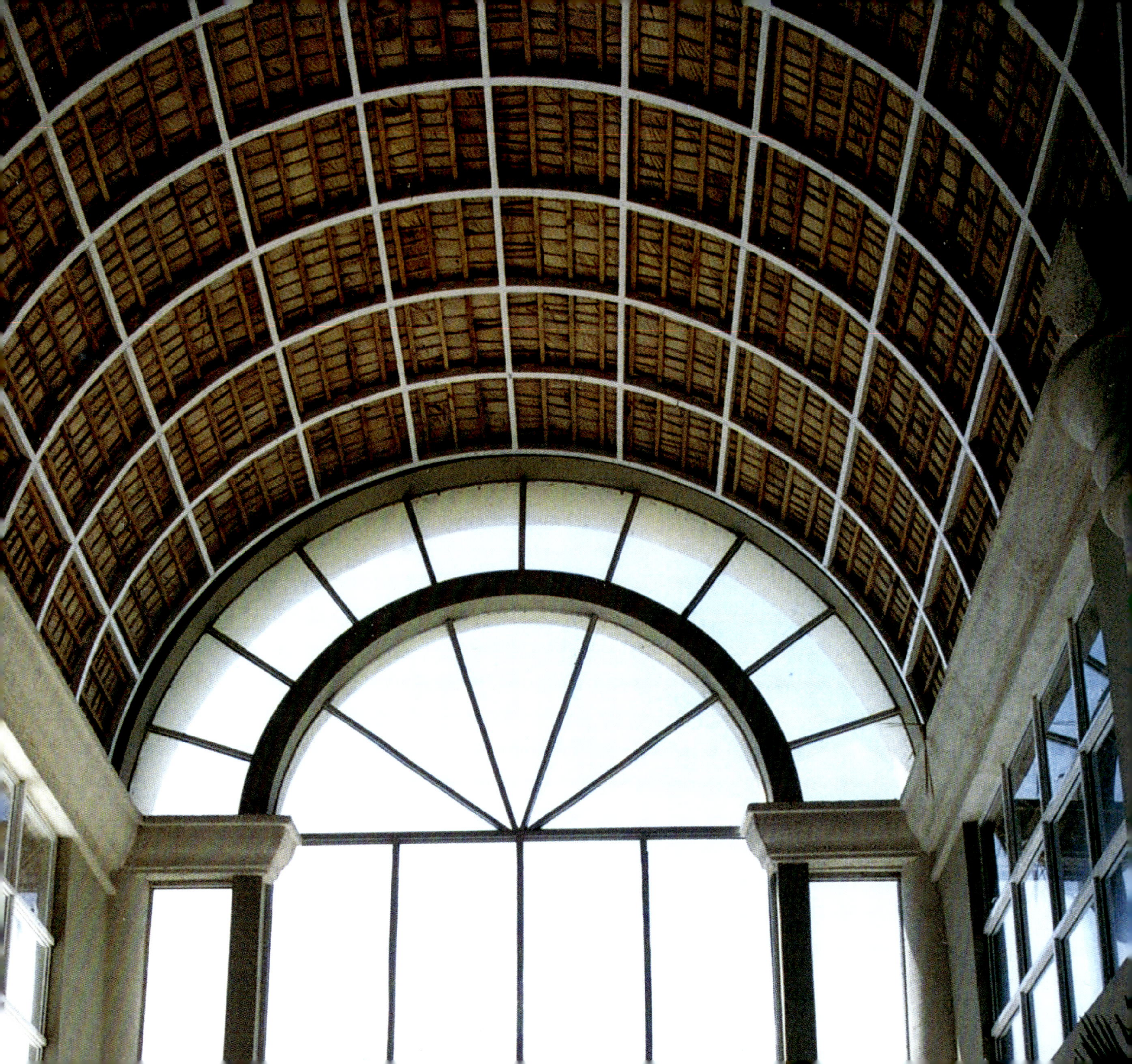

Vision into Plan/Plan Becomes Reality

The vision for Punta Cana was a simple one: an idyllic island resort in the corner of Hispaniola. But that was before Ted Kheel actually saw the land he'd acquired. Over the next three decades, the vision for Punta Cana became shaped by commerce, competition, and conservation. And, of course, by Hurricane George. The story as told by a group of disparate participants.

Before the plan there is always the vision. The vision comes before the plan. But most visions are never fully formed. They begin as half-shaped dreams, a spontaneous idea, a response to a need, a solution to a problem. As time goes on, the dream becomes sharper and more focused. What creates the focus? The focus comes from others—mentors, partners who have tried similar ideas and are willing to share their experiences. Sometimes it is trial and error. Often the clarity comes from seeing others do it right or seeing them do it wrong.

Ted Kheel and Frank Rainieri didn't have too many role models when they stumbled on to Punta Cana in the 1960s. But both were determined to make it work. In the early years Rainieri recalls juggling multiple jobs—a nightclub in Santo Domingo, an airplane ferry service—just to keep Punta Cana alive. And Kheel remembers the small cabanas, full only because he would fly his entire family there. Crunch time. Can Punta Cana ever be real? Can we support ourselves and our children through Punta Cana? Haydee Rainieri once asked her husband, Frank. She never asked the question again.

Instead, the Rainieris, the Kheels, Oscar Imbert, Oscar de la Renta, Julio Iglesias and many others came together to create a business that is so much more than a set of hotels and ocean-side residences. Grupo Punta Cana is an airport, a power utility, a real-estate development, a golf course and course-side residences and, of course, the resort itself. Around all this has grown a community of loyal and vested workers and a town, a plaza with retail stores and restaurants, and a bowling alley. Add to that the concept of conservation—a biodiversity lab that brings in scholars and researchers and creates a new focus on the environment and, possibly, a new business segment.

Tourism is a complex business. Its customers are often transient sorts with little but consumption on their mind.

Here today; gone tomorrow. And few resorts have taken the time and the effort to focus on tourism as an asset that can be rapidly depleted, especially if tourists—the buyers—and tour operators—the sellers—don't actively cooperate.

Tourism is a seasonal business. How do you sustain a community if the spending is limited to only a few seasons of the year? The future, say both Kheel and Rainieri, is in creating a year-round community—a community of tourists, homeowners and empowered locals. "The future is in real estate," says Kheel. If you can't create a year-round community, you won't have a viable economic unit.

Turning the vision into reality has taken the efforts of many—of engineers, of resort sales professionals, of Americans lured by the attractions of the DR. But everyone who has become involved has taken on additional roles in the new community.

The new community looks so much different than before. The resort still is the centerpiece. But around it is a community where people live and work. And surrounding it all is a natural laboratory that has become its own attraction, a place where scholars and researchers congregate to study and celebrate biodiversity.

Frank Rainieri

THE BUSINESS PLAN, I HAD IT IN MY HEAD. It was really a dream, not a business plan. When you don't have money you cannot have a business plan. For years we couldn't afford to pay anyone to do a master plan. The first master plan we did was in 1978 by Oscar Imbert. He was a student at that time, and he was doing it as his thesis at university so it didn't cost me a penny. I told him what I wanted to do here and there. It was a product of my personal experience of the University of Punta Cana. I didn't have anybody making master plans. Now we have beautiful master plans, but they are all subject to changes because we are constantly adjusting. One of the important things in life is that you have to adjust your business to the circumstances.

When we opened the hotel in 1987-88 we were looking for the U.S. market because it was the closest market to us, and Ted was very helpful. He knew all the press in New York. It didn't work because we didn't have the money to advertise in *The New York Times* every weekend and in *The Wall Street Journal*. Punta Cana was an unknown destination. So instead we went to Europe where they sell package tours. Europeans travel in groups, with people from their professional societies and their community. So we went there. We had an immediate response. We got the German and Italian market to come to Punta Cana. For years they were the backbone of all the development. Then five years ago, we realized that Europe was too far away. We shouldn't put all the eggs in one basket, and so we started going to the U.S. This winter there are more Americans coming to Punta Cana than from any other country. Second place is Canada and third is France. We had to adjust our business plan. For 14 years, the Europeans were the backbone. We love them and will keep them, but let's also target the North Americans. At the same time, we realized we needed golf to attract the Americans, so we began our first golf course. But the innovation was that we were going to build a sustainable golf course. We used a grass called Seashore Paspalum, a revolutionary hybrid grass that can actually be irrigated with seawater and recycled water, thereby protecting the underground aquifer. Now everybody is using it because they saw it here and it worked, but at that time it was a new grass that nobody had in the Caribbean. It means that we have been innovating.

We did the first international private airport, now everybody is doing it. In the Dominican Republic all the airports are privatized. It took us almost seven years and three governments to get the approval, and it took almost 14 years before the next one was privatized.

When we did our hotel, we set the standard: Buildings no more than four stories high or the height of a coconut tree. Up to now, with the exception of one hotel in the area, everybody has kept to the same standards that we set. We set the standard of no more than 35 rooms per hectare, which is 2.5 acres. Everybody is attracted to the low density in the area, compared with other areas of the country where there are 200 rooms per hectare.

We've been innovative: We did the first equity swap. We have done new things. It cost us more. When you are the first one you have to pay the price for it. But we have been innovative enough so everybody follows the leadership.

Adolfo Ramirez Duran

Adolfo Ramirez came to work in Punta Cana in 1996 as Punta Cana was beginning its current master plan. There was no other development except for Punta Cana, and that was a quarter of what it is today. His first task as engineer was to de-

velop Punta Cana infrastructure. In the process he got involved in the master planning of the airport.

In 2000, Ramirez left Grupo Punta Cana to start his own business but came back last year. "I came back because for me Mr. Rainieri is a leader, and when you are in a relationship with a leader, sometimes they convince you and you come back," says Ramirez. "Mr. Rainieri has always had a very clear vision of what Punta Cana and the whole area is to be, in terms of the tourist and the type of services we are offering, and I believe it has become part of my own life."

WE BEGAN TO IMPLEMENT THE MASTER PLAN in 2003, with new personnel. In the last two to three years, Grupo Punta Cana has grown from C to B and we are planning to go to stage A. We are building up to that. We can build the best facilities, but if there is no one to manage correctly, then it won't be any good. So we contracted out to several companies to help us in the master planning, putting together the ideas of Mr. Kheel, Mr. De la Renta, and Mr. Iglesias. We are executing the master plan with people and companies who have experience, and have developed similar projects out in the world.

We produce our own energy, our own water. We are now contracting out garbage collection. Up to 2002, we were doing the garbage collection ourselves. Now we are using another company that is established in the area. They can do it better, so we gave the contract to them. We believe that by contracting out key components of our plan to other companies, we can all share in the growth. We truly believe that we cannot be an island in the area. The whole process of organization and development has to benefit everyone. We cannot finish a project and feel like we have done a good job if the nearby towns don't get water, don't have sewage treatment, or can't have their children educated. We are opening a school in Veron, a high school, because we believe in education. If we have that, then we will have a better labor force. More for everybody.

We are also planning and developing a second golf course for this year. In fact, we plan to build five to six golf courses in the land that we own. We also are planning a polo ranch and other activities. We believe this is complementary to the whole package we have to offer. We are planning to include the services we are doing in the ecological center and in the biodiversity center. We are talking to several universities in order to establish a campus in Punta Cana. We have to include marine biology. If we sell a beach property, we have to sell a good beach. So we are working together with several universities and consulting companies who have devoted their lives to working on the subject, in order to improve and protect the marine environment and land environment.

The reef is a lot more complicated because it depends on the natural forces and the impact of man. All over the world, reefs have been degraded. And though the picture is starting to improve, it will take time. It is not something you do in three to five years; it takes more like twenty to fifty years. There have been several mistakes in trying to improve reefs. When you try to do something very complicated in the environment, you have to be sure about what you are trying to do. Sometimes it may be better not to do anything at all and improve those things in the environment that can be improved. We are working with the best in the world, so we hope we will get the results we are targeting.

Punta Cana has always had a master plan, but on different levels. Because the Rainieris initially didn't have the money to pay a consulting company, they did it themselves. The first master plan, when I arrived, was the master plan born in 1978, when they began construction of the hotel. With that master plan, they were planning two golf courses, a marina, a social area, a commercial center, and an airport. They say that

Punta Cana, of all the projects I know in DR, up till now, began with good planning. It may be the only tourist place that has a master plan, infrastructure, and services to support that infrastructure.

When I arrived the first concern was that we had to provide basic services. The energy was supplied by three 500 megawatt plants at the hotel, and one 500 kilowatt plant at the airport. The first thing we did was to plan for energy. Now we have EMB plants, five and ten megawatts and we signed a contract for the next plant that is going to be a 6.5 megawatt with co-generation. We are producing our own energy from fuel oil. We may be able to do a wind energy plant, but we don't have constant wind. A solar experiment didn't work, and the technology is very expensive.

We look at the master plan and the development process as a whole process. At first we had to provide all the

Vision into Plan/
Plan Becomes Reality

services because we were isolated from everything else. In the beginning, the road to Punta Cana was four or five hours and when it rained you couldn't come here. So Punta Cana had to develop a type B road. The government didn't contribute anything. We had to develop our own water and the waste-management program. Everything has been growing together, the services and the infrastructures. When we began planning for the golf course we improved our wastewater-management program so we could irrigate the course with the recycled water.

Over the years we have been charting a good direction when it comes to getting professional help. We have top supervisors that are already experienced so that helps us develop downstream. We don't build ourselves, we have a bid process, and we have a roster of companies, depending on the job we have to do. We have selected four or five different companies for different types of jobs. So, for example, when we build a ramp at the airport we have five or six companies that we use, so they already have the experience to do the thing,

When it comes to the general work force that's a different thing, because most of the work force is not trained. So we have a good team of supervisors to provide for that lack of training, for the work force at the bottom. We are beginning to develop some training in our own personnel, and we are improving our own personnel. When we achieve that, we believe things will be easier. At the same time you improve your own personnel, and when you work with somebody else, they indirectly will train themselves.

When I began working here in 1996 we were doing some stone work. The guy we picked to do the job at that time came to work on a motorcycle, and he didn't have the tools to do the job, so the work was very manual. Over the years, the same guy is working as a contractor. He now owns four pickups, all the tools, and what he was doing manually at that time he now has several workers to do it for him. He has become an entrepreneur. And that happens in almost all areas. Because when you take care of people, and you improve the way they do things, and their life, most of the time they will succeed. Sometimes a contractor may not be the least expensive to do the job but we continue to use him, because he has the training, he has already done it, and we are sure of the quality of the job he will do for us.

Our growth depends on our ability to work with smaller partners in that growth. We have helped to develop at least twenty new small companies and over one hundred companies directly and indirectly.

Is our planning perfect? No. Lots of things can go wrong. First, the market may change, and we may not be able to execute on our plan. A good example is 2001. Indirectly we received the impact of the problems with terrorism in New York.

We have already made some mistakes and have tried to learn from them. If we make the wrong decision it will affect the planning. For example, if we build something in the wrong place, we can make the growth of certain areas more difficult, or impact the environment, or some other people's lives.

One thing we have tried to do without fail is to maintain respect for nature. We try to do everything to respect nature, and protect it as much as we can.

Sure there is a cost. We justify it because we believe in that. And we believe the trees will take twenty years to grow in this kind of environment, so we try to preserve those trees, not take everything down. It has a cost but at the end it pays. It would be easy for us to take all the garbage and dump it in Veron, for example, but it will affect those people. So we decided to pay a little bit more to a company that will take the garbage and have it disposed in the right way. We are forcing them to dispose it in the right way. We are asking them whether they have the license from the municipal department. And the license means that they are being supervised.

We can make mistakes, and we will make mistakes. I have always said the goal is to correct the mistakes on time. Then you are doing something right.

When we are thinking of developing something new, or remodel, we interact with all the departments from the financial point of view, from the management point of view. About a year and a half ago, for example, we began to develop a program to protect the beaches. That project was approved from the financial point of view, but all of the departments then had to be involved because that is the process.

We benefit a lot from the input of people like Oscar de la Renta and Julio Iglesias. Although the general master plan is unchanged, the implementation has improved, because of the experience and insights they have provided the company. Their ideas and opinions are important because they have gone to twice as many places as we have. When you see and live in other places, and see the type of services you will get from Hotel Mandarin Bali, then you can come here and import new ideas that we can use to improve the operation or the ecological results. It's like a sandwich. Before we had one part, now we have both. They have been a great asset to the way the company thinks, and has developed, keeping the core philosophy in mind but improving the way things are being done.

Going forward, the biggest challenge is not building more; it is creating the structure that the company needs to have, to manage the process, now and in the future. That involves developing human resources. When you have the right human resources, you will be able to develop the structure, and build whatever you want. The company has to finish forming the structure that allows the company to develop. The biggest challenge we have is to improve our own human resources—everything else is important, but is collaborative and will depend on that. If we have good human resources that can manage the whole process of growing and grow together, then we can be successful.

I would say that we have more than adequate resources in the Dominican Republic, but whenever we feel that we have a specialized need, we go to the U.S. or Europe and hire people. The point is that when a company is growing too quickly, sometimes the structures lag behind. You can't let the gap grow too big because then the job becomes impossible. That is the biggest challenge: to keep the human resources growing. From the guy who sells to you at the restaurant, to the club manager. The whole process. The job is not owned by one individual, it's a collaborative effort for the group, and all of them from the bottom to the top have the same feeling about the company, the same feelings of where we are, very clear feelings of where we are going, and how we are going to achieve that.

Oscar Imbert

Oscar Imbert's father is best-known for participating in the assassination of dictator Rafael Trujillo and being the only one who survived the mass slaughter that Trujillo's sons then carried out. Oscar, the son has made his mark as an architect who has focused on melding traditional with the modern. At Punta Cana, for example, Oscar Imbert, has deftly blended tropical Dominican, Spanish and native Arawak Indian designs into a unified series of low rising structures that neatly reflect the past and present.

Imbert has been involved in the planning and development of the Punta Cana from the very beginning and much of the credit for the uniqueness of the resort has to be given to Imbert and his persistence in melding the past and the present.

I BEGAN TO THINK ABOUT WORKING WITH thatched roof when I was studying architecture at the university. In this country, the tradition of thatching and the thatched roof almost disappeared from the 1930s to the 1950s. I proposed to my advisor that I would like to do such a project because the roofs were very attached to the history of the Taino Indians who were the original inhabitants of this island. They had different types of small structures that they call canae.

When the Spanish arrived here they wrote in their books that some canae can hold up to seventy-five Indians and the cacique. No matter if they were Indians or Spanish—seventy-five men were seventy-five men.

So I designed that hotel in the place where the Punta Cana Beach and Resort is now. And it was great. With that project I qualified as an architect. So I became an architect designing a project in Punta Cana. After working for one year I went to New York to pursue my master's degree at Pratt Institute. I proposed to design a marina where that little bay next to the hotel is. I had several photographs of the area and I had my '50s project with me. I always mentioned at Pratt that I was very concerned about nature and the plants and so forth. When my third presentation to the jury came and I told the jury about my concern about plants and coconuts and they didn't let me talk too much. One of the professors told me: "Listen, I think you should go and work again on your project." And I said why? And he said, "If you are so concerned about the coconut trees, how come you are going to make a plaza that will cut down all the coconut trees. Your concept isn't very clear." I went back to my drawing table and started again.

I selected a lot in Puerto Plata in the North Coast and designed that hotel for my thesis at Pratt. After getting my master's thesis at Pratt, I came down and showed that project to the developer and he decided to build the project. That design by that time was very, very clean. And that hotel was the first hotel in the North Coast to have a thatched roof. It was very Victorian. I tried to re-create the old Victorian architecture of the island with a new look and a new vocabulary. Then I got the chance to come to Punta Cana and help Frank with the airport. Frank called me one day and said, "Listen, Oscar, I need you to design the airport. I don't have any money to pay for the plan. As always, if you want to be in this you are welcome to join in."

I came in with the idea of designing the airport with a thatched roof and using the wood that we cut during the construction of the airport runway and using the coral rocks that you can find anyplace here. We always liked the idea of Punta Cana—which stands for Point Cana—because here there were a lot of Cana trees just growing over there. So I thought it was a good idea to use architecture with very natural elements and that was the idea behind the airport with a very organic architecture. The first building was really very successful. A lot of people thought it was very convenient for tourists because when you come to the Caribbean you really want a building that welcomes you. We started on the right foot. At that time,

VISION INTO PLAN/
PLAN BECOMES REALITY

it was really fresh. Most of the year you can come and see how the breeze blows from one side to another and how the Cana roof really works. The Indians were not stupid. They were aware that the Cana would stand up to hurricanes, to water. The Caribbean image needs to have an overhang of Cana. As soon as you see the fingers dancing in the wind, you know you are in the Caribbean.

I began a master plan for the Punta Cana property to build a development based on low density and trying to respect all of the natural beauty of the site. After the first project we started looking for the nicer spots. For the first master plan we decided to keep the ecological reserve, which consists of fresh water springs. Next is the biodiversity lab. It is incredible how scientists from various parts of the world come here to study. Even though this is an island we have the highest mountain in the Caribbean. Here you can get a variety of plants and a different kind of resort. It's not too common that you can go to a resort and find a biodiversity lab and such things as DNA sequencers. It's a unique thing that we can show the tourists from different parts of the world. That opens other levels of sightseeing.

When I designed the Punta Cana beachfront I designed five different floor plans for the small beaches and I went into the coconut grove and found the templates for house sites that used the least coconut trees. I chose those templates. Actually I was lucky in a sense. At the beginning of construction lightning hit the coconut grove and took 15 trees. And if I did seven houses taking only seven trees, I was in good shape. By that time I always fought for the right of a plant to survive. Here we have some very old guajacans, some very old hardwood. Why should I be allowed to take such a tree down just because I am a human and can take it down? I don't believe in that way to go in architecture. We have a unique island and as an architect I feel very responsible to know the inhabitants of that spot. So if you see the plants and see how they sway and bend with the wind you can understand the intensity of the wind. But if you come in with a bulldozer and mow down the trees you'll never understand the wind. Similarly with termites: If you see termites in an area you know there can't be floods in the area because termites are very intelligent and choose dry land for building their nests. So you have to pay attention to all the detail and see how the sun comes and how the full moon is very spectacular. If you pay attention to all this you can make an architecture that is unique to the essence of Punta Cana. I think most of my work has been along the same lines and has the same common elements since 30 years ago when we decided to build the first hotel. I decided to create holes in the overhang to let the palms cross by. That doesn't offend anybody, not even Frank Lloyd Wright. The way we deal with nature and the integration of the landscape here has been present from the beginning and has continued. Even the airport is a thatched building with a thatched roof. We had to mix

Vision into Plan/
Plan Becomes Reality

38

some material to shield it from a hurricane. But the airport still has the feeling of a building in the Caribbean. I am really encouraged when I see people who come to the airport taking pictures from every spot possible. It tells me they are happy. It is a unique airport and I haven't seen any airport like that in any part of the world I have traveled to. People feel relaxed when they arrive. They feel okay and they relax. So I like that.

There are a lot of developers who see a project and see only money. When you are trying to design an environmentally friendly project it is only natural that to preserve nature you have to give to nature. You have to give something to nature especially if you want to protect it. In a high-density project, if you take up every square foot of the environment you are going to lose. When people take up every ounce of air it affects the neighbors. Trees don't give up only fruits, they give shadows. People everywhere now are getting more concerned about protecting the planet and that is a very good thing.

Richard Aurelio

Richard Aurelio served as deputy mayor to New York Mayor John Lindsay and as New York Senator Jacob Javits' press secretary. He then joined the fledgling cable industry to become president of Time Warner's New York City Cable Group. After almost 30 years in cable he retired and moved to Long Island.

JUST BEFORE I RETIRED FROM TIME WARNER, my wife and I were looking around for a winter home in the Caribbean. We looked at a lot of them—Barbados, St. Bart's, Martinique. You name it, we covered it. Then somebody told me about Ted Kheel's project in Punta Cana. I knew Ted from the days when I was in city government and he was a labor mediator. So I called him and he told me that they were beginning to sell properties. He was very low-key. "Go and take a look," he said.

When we arrived in Punta Cana, we fell in love with the airport, thatched roof and all. On our travels to Mexico we had really gotten to love houses with thatched roofs, so when we saw the same at the airport I asked who the architect was. They told me it was Oscar Imbert. So I called Ted Kheel and said I would be interested in building a house there, but only if the architect was Oscar Imbert. And the design would have to include a thatched roof. The house was built in 12 months.

After we bought the property we had the worst hurricane in history. But instead of building in September 1998 when the hurricane came, we began building a month after. The hurricane didn't scare us. Chances are there won't be a hurricane for the next thirty to forty years

We also fell in love with the people. Everybody here is trying to help build a community and it's also becoming international. It's pleasant to go to Punta Cana from New York and not feel the sense of being in another place. Some of the other Caribbean islands are so anti-community. As we have started living here there's also been the growth of an economic community. Stores have been cropping up in response to demand. There was a time we had to go to La Romana to shop or get things shipped from Miami. We now live our lives here.

Even my wife has become entrepreneurial. Some of her friends had started a store in the plaza. And my wife was interested in selling resort-wear. So the friends gave her a corner of their store to sell the material. Last month, they called her to say that the entire lot had been sold and they needed another shipment.

We are living in a community that really cares about conservation. We have seen other Caribbean communities where the idea of conservation is given lip service, but here they see it as an integral part of what they do and what they offer.

Michael Davison

A veteran in the business of selling resort real estate in the United States, Michael Davison moved to Punta Cana in 2002 to help Grupo Punta Cana build a sales organization, train the sales people and help them sell real estate around the marina, the golf courses and Corales. The move worked to boost sales as well. "When I came here April a year ago they had made six sales. From April to December we did 50 and then up to April we did another 35. We are almost running out of inventory," says Davison.

Davison arrived at a time when Grupo Punta Cana decided to break ranks with most of the resorts in the area by switching from an "all-inclusive" vacation plan in which tourists committed to one fee that covered everything from food to lodging, to a plan in which all services were separately paid for. The change, in the long run, promises larger revenues for the resort but it also deprives tour operators of the control that they like to have. For a time, a number of tour operators decided to steer their clients to other resorts.

WHEN I FIRST CAME, THERE REALLY WASN'T a specified plan on how to build a residential community. They were selling home sites for cash. When I looked at the manner in which they positioned the sites, much of it didn't make sense. On the front nine of the golf course—which they call Tortuga—they had all of their home sites on the market. From a real-estate point of view that's not a good thing to do. You want to come up with a plan that will use the real-estate sales to help build up the entire community and to enhance the value of the property as well. The sales aspect of this is only one thing. It's all the pieces that are put together that make the sale effective.

There is now an understanding of where we are all going. Grupo Punta Cana knows the market it wants and I am helping them design the product that will appeal to that market. All of the pieces have to fit together because what we are selling is a lifestyle.

We want lower density. We want to be family oriented. We're going after professional and well-to-do people who want a vacation home or a retirement area or even a primary home. Those types of owners will be here. Right now we're predominantly vacation. But Americans who are buying their second, third, or fourth home and developers from the States are looking at the Caribbean because the great places and the unique places on the mainland are gone. You can't afford it or you can't get it. Some of the best lands in the U.S. are owned by the government—they are national parks or state parks and the government will not cut them loose for developers. There would be a revolt if they did. That is one of the reasons why Panama, Costa Rica, Puerto Rico and the Dominican Republic are now so much in demand.

Frank Rainieri was the original visionary for this place. It was all jungle when he was helping Ted Kheel set up in the Dominican Republic. It started out with trader Frank who said he was going to make Punta Cana a major Caribbean resort destination and they vilified him. Now a lot of people are eating Dominican crow. He sees beyond what a lot of developers see. He is working on commercial and residential areas for the people who work here. He is talking about workers owning homes or greatly reducing the cost of homes. It's a very rural area. Some of the people have had very little education. People lived off the land or were fishermen. He brought a school here; he brought a church here. He rewards people who take a job here and who know more than one language. He's a businessman but it's not all just profit-making.

In the long run, the two major income producers to keep this place viable will be the airport and the resort. At the airport they are bringing in more than a million people at $10 a head. I would assume that there are also rental fees off the top and a piece of the gross, as well as fees for fuel and services from the airport.

When they switched from an all-inclusive plan, occupancy dropped substantially because the tour operators in Europe and the U.S. like all-inclusive, and we were the only hotel in the Punta Cana that wouldn't go along. They were punishing us. Now occupancy is coming up again and we are finding revenues are better because we are getting the more-affluent guests and they are spending more money. It's changing.

Vision into Plan/
Plan Becomes Reality

We are getting a lot more business. We don't have 450 rooms filled all the time but the occupancy rate is a lot healthier than under all-inclusive.

How do we market ourselves? We are not going to appear on TV as a spring-break destination. And we don't call ourselves an eco-resort in the manner many others do. But we pay a lot of attention and regard to the environmental aspects. The residential areas, for example, are designed to use golf carts and bicycles. For the golf courses, they are using a special grass that can be watered by seawater and only uses 50% of the insecticides typically used.

A lot of people pay lip service to environment. A lot of the times environmental assets take a back seat to the environment. But if you travel, you'll see Punta Cana hasn't done to the environment what many others have done. We are doing environmental mapping and one of the reasons why we have been slow is because there are wetlands there. They don't have the trees that we have in South Carolina but they do take into account the need for indigenous flora. There is a lot of regard and respect here for nature.

When we get ready to go into an area we ask: Are we eliminating a rare species? Are we destroying the wetlands? In this area it would be very easy to put in some lots. But there is a freshwater stream here and we just won't touch it. In terms of environmental concerns, they are at the top at this place. Punta Cana walks the walk and talks the talk.

Three to five years from now, we are looking at a community of four hundred to five hundred homes. We'll have three golf courses, possibly four, in addition to this one and it will be a strictly real-estate deal. All of the golf courses will be inland—it won't be as exciting, but it will enhance the value of the interior. When this is all done the legacy will be a city.

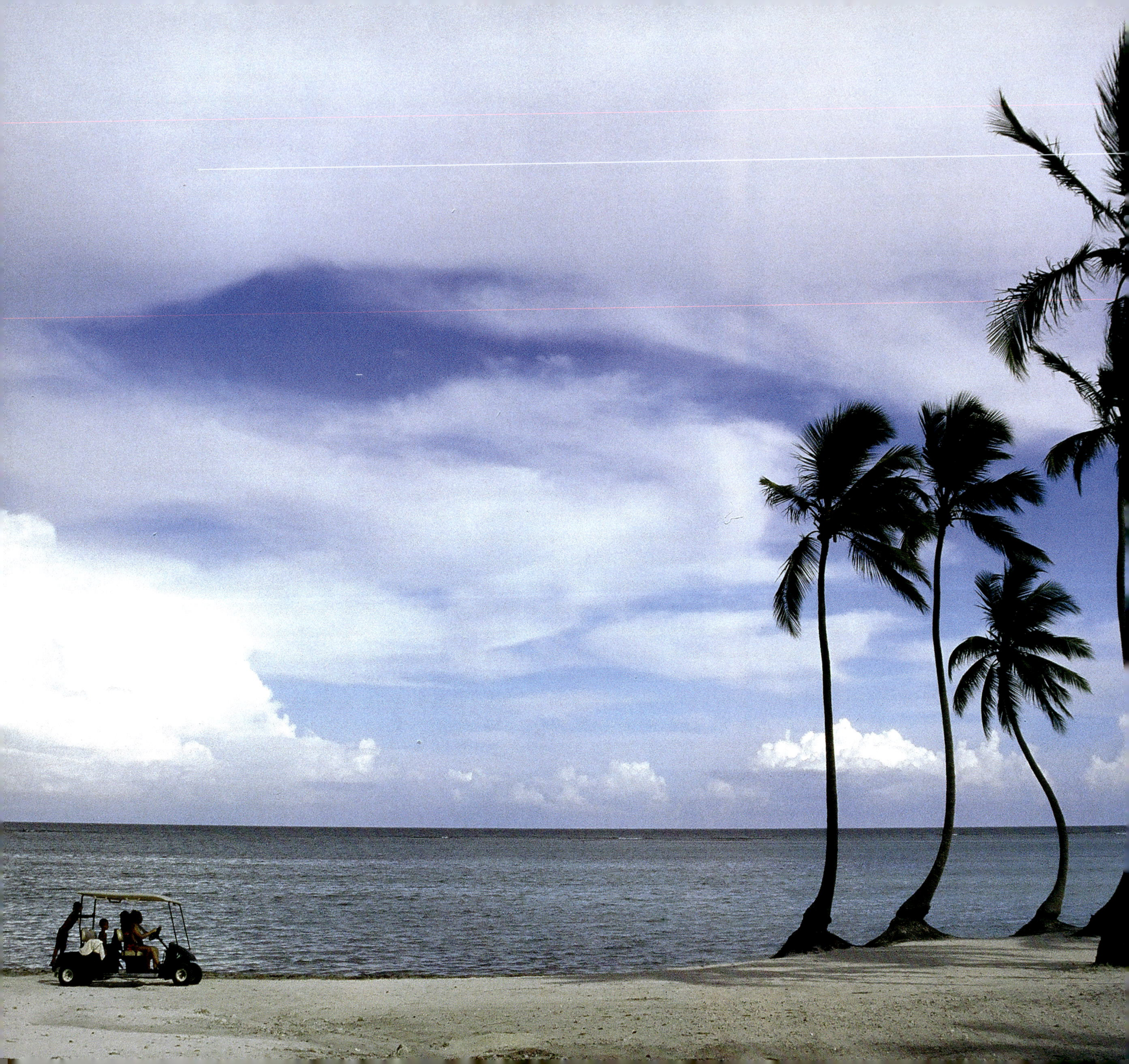

Conserving Nature/Building Human Capital

The idea that tourism and biodiversity are related is not unique. Eco-tourism, after all, encourages travelers to visit the remote reaches of the planet to explore the diversity of plants, insects, and animals. At Punta Cana, the idea of biodiversity began somewhat simply—as a means of creating greater consciousness about the implications of development on the local environment.

Kheel, a mediator and expert on conflict resolution, felt that understanding biodiversity in the Dominican Republic would greatly assist tourism and the environment.

Ted Kheel:

Tourism depends so heavily on nature. People go on tours because they want to get to a warmer climate; they want to get to more sunshine, to beautiful landscapes, to beautiful waters. Tourism depends very heavily on the protection of the environment of the places that attract tourists. For tourism, the environment is the attraction. If the environment is damaged, the business is damaged. Tour companies have a tremendous stake in the protection of the environment. It's just like a company that has a plant. If they don't maintain the plant, they don't take care of the machinery, the plant becomes inefficient unworkable. In the case of tourism, it's the attraction itself that needs to be protected. There are few businesses that I can think of where the protection of the environment is more important for the tour operator than tourism.

The Punta Cana Center for Sustainability and Biodiversity has become the means for Grupo Punto Cana to examine itself and monitor the impact of its development. But as it began to attract universities and university researchers its stature grew dramatically. Previously. Santo Domingo's Jardin Botanics Nacional had occupied the researchers' attention, but the Punta Cana biodiversity laboratory was now carving out its own niche. It was easy to get to: Punta Cana airport was only a stone's throw away. The living conditions were comparatively luxurious. But most of all, the lab and the resort together provided a unique combination—a means of testing the real-

Pedro Rodriguez is the Ecologocal Foundation Coordinator. Inside the biodiversity lab.

Conserving
Nature/Building
Human Capital

ity of conservation in the marketplace. And with Ted Kheel's ability to lure some of the U.S.'s most respected scientists for visits and research stints, the Punta Cana Center has achieved credibility in a remarkably short space of time.

There aren't too many places where tourism entrepreneurs are willing to establish conservation and ecological responsibility as a cornerstone of their own development. But the Rainieris' willingness to do so has given tourists to Punta Cana a new dimension to explore and an additional reason to visit.

The Punta Cana Center continues to evolve into a dynamic center for learning. Dominicans now study their environment side by side with scholars from Cornell, Harvard, and Columbia. An evening stroll leads to a lecture by Harvard's Brian Farrell on insect life in the Dominican Republic. He talks of how he and E.O. Wilson, the former Harvard curator of entomology, teamed up to solve the mystery of a sixteenth-century ant plague that forced the relocation of Santo Domingo. And he describes how he has been modeling a bioinformatics database of Dominican insects that will be accessible and searchable from anywhere in the world via the Internet.

There is also an elementary school and now a polytechnic school. The desire to provide education for resort employees and their families is not altruism. With government-run schools in a poor state, Frank Rainieri knows that his employees want their children to be educated and that many might leave if educational facilities are not easily available. There was a time when most of the Grupo Punta Cana employees lived in Higuey, the nearest large town, and commuted to work at Punta Cana. Rainieri had a work force, but not a community. The Punta Cana Center, the elementary school and most recently the polytechnic have created an educational environment, a place where employees send their children and go themselves. It is an opportunity not simply to be educated, but to lay the foundation for a different life--with Punta Cana or outside of it.

The Center and the schools have also become the backdrops for a number of other initiatives designed to create a more informed community. There are women's groups being formed, adult-education classes being offered, and health-care information projects established to talk about AIDS. And there is a new effort to expand the library to promote literacy in a country where secondary school enrollment is ranked 10th among 13 Latin American countries.

The schools and the biodiversity lab are part of a new

way of thinking about community. It isn't enough to provide a job. Beyond the job is the desire for self-improvement and for the caring and well-being of the family. Unless one can create an environment in which the worker and the family flourish, the task of business will not be complete.

Frank Rainieri on Grupo Punta Cana's education initiatives:

THE STRATEGY IS TO BE A MODEL, because sooner or later the world outside will have to adjust itself. Yesterday, I was speaking to the director of the school. Three years ago we had 158 children. Two years ago, 225. Last year we had 260 children in the school. Now we have 315 and we are soon going to have between 330 and 350. Some of the grades are closed because we don't have space. In four years we have more than doubled and people are paying to go to school. We give scholarships. but not to everybody. It shows you how fast change can happen.

It's a changing mentality. People are sending their children, looking for good schools. Before, the attitude was, well, send them to school because we have to send them. Now they're thinking, let's send them to best schools, and they are realizing that this is the best school in the area.

In the whole of DR there are probably 20 public libraries open for eight million people. There are maybe three or four, for sure, no more than five that have more than 5,000 books. Our next project is to build a library. We're working on the building already. It's in the budget for fiscal 2005. We'll have it ready for September 2005. State of the art. I'm going to show people it can be done and it will be used. Probably the first year there will be days when there will be no one there. It doesn't matter. After three years, we will be too small for the area.

You have to lead by example, not by telling people. People in the Third World are tired of taking orders. They have been obedient for centuries. "Don't do," is the key word in our countries. So we have to show them what they can do, not what they cannot do. That's why I never talk only about sustainable development in nature. If people don't develop socially they'll destroy whatever we try to do. We know it. We saw our forests being destroyed to make charcoal for years.

Entrance to the elementary school.

Then we have to show people, work with them to improve their quality of life, but more importantly their capacity to understand why certain things can be done and certain things cannot be done. If someone is 25 years old and never went to school, how can you teach that person what to do and not to do. Only by example.

I believe in sustainable development, sustainable tourism, and sustainability. Why? Because we still have a million people who go to bed with one meal a day. You know you have to try to make things perfect, but you also have to realize that you can't make everything perfect. Some issues you don't fight. You have fishermen who only know how to fish. Unless you can provide them with a fishing job, and a fishing alternative, they'll keep on doing it. How much can we preserve? How can we use what we preserve? What I have tried to do is make these two concerns work for the benefit of the people. That's why at the same time that we preserve nature and we have a biodiversity center, we also have a school. We're building another school for the people, because if we spend all the money preserving nature and we don't put the money to educating people, we're just wasting our time. Because when people are hungry, or when they are not educated, there is no way that they are going to preserve nature.

When the project began there was only the economic aspect of things, now there is the social, and a biodiversity aspect that you have to take into consideration when making a business decision. Three concerns have to be taken into consideration.

When I was seven or eight years old, in the 1950s, my mother and one of my aunts organized a group of women. The aunt worked with the kids with polio. My mother worked with the kids with TB. At that time, people wouldn't get close to people with TB. Every month my mother would take a group of people to entertain them, and my brothers and I were the ones who would pass around to get food and other things for the kids. I learned very early in life that there are two things—the material and the spiritual. The material is one that is wonderful and everybody wants. But there are spiritual things that are more important. More than that, we have the privilege of life, and if we are not concerned with other people, who is going to be concerned. So I have worked all my life through the Catholic Church with the poorest neighborhood of DR.

In 1972, we opened our first little school. We had 18 children. Thank God I always had the backing of Ted, who has a social conscience. I always had a dream, that someday Punta Cana would have schools, hospital, houses, because I always believed that my children and their children should have the same options in life. That is why I have always been very involved. I did the same thing my parents did to me, when my children were small, I took them to the poorest neighborhood of DR, and I let them play with the poorest kids. Today we are doing it indirectly. In a society like ours, it's not like U.S. or European society where the poor children go to the public school with the wealthy children. The children here go to public school and they don't make any standard and the middle and upper class and wealthy never go close to one of those schools. In the school we've built over here, you'll see the children of my chauffeur, and the maids of my house—they pay according to what they make—attend school with my grandchildren. My chauffeur pays $1 per child per month. My son-in law pays $75 per child.

In the beginning, even my wife was concerned—and, incidentally, my wife was a rebel when she was young. But it's worked out beautifully. The standards are very high. For the second time, our school got the first place in the national math tests of DR.

Ligia Henriquez

Ligia Henriquez' official title is Directora Centro Educativo Punta Cana. Over the last two years she has been principal of the school at Punta Cana, a school that provides six grades of education to the children of the area. The school is a private sector project, financed in part by the Punta Cana foundation and in part by the fees that are charged for admission. With more than 300 children enrolled, the school is unique in its attempt to educate everyon—children of senior executives and their employees—all within the same walls. There are a few local schools funded by the government but they suffer from a perennial lack of funds and resources and, most of all, books. This school, with its hard-acquired computers and books, is a surprising departure and a remarkable testament to the resort's involvement with its community.

WE WANT TO HAVE A SCHOOL HERE that offers quality education and will fulfill all the needs of area. The school is not just for the children of the employees, the school is of the community—all the people who live around the school, even the people who work in other hotels. We are now a multicultural school and we play an important part in bringing people to this area by assuring them that their children will be safe and have a good education.

The children of the Punta Cana employees who attend the school pay according to their parents' salaries. You can have families who pay 145 pesos monthly and you can have parents who pay 2,000 pesos monthly. That's not what the cost of the education is. Our school is partly financed by a foundation.

For people who are not employed here, the tuition is higher. However, we also try to provide special treatment to the children whose parents are not employees. We have a special pickup location for them. And although the minimum they pay is more, they are provided a quality of education that they can't get elsewhere.

We also have an after-school program. We ask parents whose children need more work to have their children attend. We know that some children lack knowledge in some areas and we help them after school. We have a program on how to study and we help them to perform. The classrooms are not that big, so that helps a lot with providing children with more individualized attention.

We are starting a mentoring program. In order to graduate, our students need 60 hours of community work. Some of the students do their community service by working with children who are not as gifted as they are. There are children whose parents are illiterate. How can they to learn at home?

Ligia Henriquez in front of a student-painted mural.

CONSERVING NATURE/BUILDING HUMAN CAPITAL

Shown here under consrtuction, the Punta Cana Politecnino (High School) opened in the fall of 2004.

Well that's where the mentoring program comes in. For example, children who can read English help others read.

University students who come to Punta Cana for the biodiversity labs also come in and do seminars and lectures. They provide examples and information that we normally cannot get.

Education is something very important. Children are some of the best ambassadors when it comes to spreading the word about the environment, especially within their families. We want the children to be creative. What is a garbage can? That's where families can put their garbage. Why not throw it in the street? It's the kids who are there to explain to parents why garbage shouldn't be thrown out onto the streets.

We also have a program of continuing education that is there to help adults enhance their skill base and experience. Some of the subjects are in the skills they need at work, such as English, French, project-management techniques. These are the kinds of seminars, lectures and workshops that we offer. I am hoping that this program will become the cornerstone of learning in this area.

The public schools in the country don't have the resources to provide high-quality education. And although we have more programs than other schools, our resources are limited too. We put a lot of faith in reading and using the library to expand the children's knowledge of the world. If we don't have enough books, we encourage the employees to provide the books that others can read. Books in English are very expensive. So we need a godfather or a godmother who will donate the books and the books will stay on after the student has gone.

Maybe eventually we will have a university here in Punta Cana. If not a university, then a technical institute, a place where people can train for service careers.

Kelly Robinson

Until early 2005, Kelly Robinson, was director for Environmental and Social affairs of the Fundacion Ecologica Punta Cana. The Ecological Foundation is the owner and manager of the Punta Cana Center for Biodiversity and Sustainability. It manages the Indian Eyes Ecological Park, the organic fruit tree and vegetable garden, the petting zoo, and a series of Nature Trails. As director she also helped to administer all the operation departments which is the pesticides, all the trash pickup from the complex, the corporate nurseries that produce all of the plants and organic goods for the salad bar. The Foundation also conducts all of the testing of coastal water, treated water, potable water, and all of the environmental permits the company as a whole is supposed to have.

I HAVE BEEN ASSOCIATED WITH PUNTA CANA for 12 years now. I had originally come to the Dominican Republic as a Peace Corps volunteer. My background is in hotel and tourist management. That was what I studied and graduated in. I always dreamed of living and traveling abroad. My parents instilled that in me. I always had an affinity for Latin America. I was hired right out of college by Hilton and they told me if I didn't work five years domestically I wouldn't be able to go abroad. And I thought, My God! If I stayed five more years in Texas I would never leave. My parents had been Peace Corps volunteers, so I applied, and to my great surprise was recruited. I say surprised because they take one out of ten people that actually apply for the job. When they first assigned me to the Dominican Republic, I had no idea where the DR was. They sent me down here and I was an education volunteer for two years.

I worked as a grass-roots volunteer in a very small community. The community said they had 800 people but they must have been counting the goats and the chickens or something because there were only 300 people that lived there. Dominicans are very hospitable people, very kind but also very poor, especially in the rural areas. For somebody coming from mid- to upper-class America, it was amazing to me how generous people were. Because I worked in a school it would always pain me to walk down the street because everybody would want to me to eat with them. They would literally share what was on their plate for the day with me even though they had no more than what they were giving out.

It was a very humbling experience and I think that my time in the Peace Corps prepared me for a professional career and my success in the Dominican Republic. Toward the end of my term—the Peace Corps is a two-year term, one of my program directors invited me to a United States Information Service forum on eco-tourism, knowing that I had been a tourism major and thought perhaps some linkages could be made there

It was there that I saw one of my first IMAX films, one of the Amazon. They had shown a part where they were burning the Amazon forests. You could see it from outer space. It entered my head that somehow I wanted to be in conservation and tourism. At that time, the word eco-tourism hadn't been coined, but I knew it was something I wanted to do. And it wasn't that I actively pursued it, but it's been one of those things that my higher power, god or whatever you want to call him or her, really intervened on my behalf and allowed me to pursue a career here.

I went to this forum where Mr. Rainieri was a speaker. At that time Mr. Rainieri's brother was the minister of Tourism. I was living in the countryside where there were only 300 people. The place had no electricity, no running water, there was one road. The extent of my life was how was I going to get gas for my stove that had no gas. So I came in and I heard Mr. Rainieri speaking about how he has 450 rooms. At that time eco-tourism was very narrowly defined. It still is narrowly defined. I got up and asked some very straightforward questions. I didn't care; I was a Peace Corps volunteer. I already had a job. And during the question-and-answer sessions I asked Mr. Rainieri about prostitution and environmental degradation and how could he pretend that a hotel with 450 rooms was an eco-tourism resort. He was a wonderful speaker and a very charming man and he deftly worked his way through that. But after the presentation he came down and said who

CONSERVING NATURE/BUILDING HUMAN CAPITAL

PUNTACANA
RESORT & CLUB

GRUPO PUNTA CANA INAUGURATES THE ANN AND TED KHEEL POLYTECHNIC SCHOOL AT PUNTA CANA, DOMINICAN REPUBLIC

Punta Cana, Dominican Republic (Nov. 4, 2004) – Grupo Punta Cana, the major developer of Punta Cana, and the exclusive Punta Cana Resort and Club, recently inaugurated the Ann and Ted Kheel Polytechnic School in tribute to the legendary New York labor mediator, Theodore "Ted" W. Kheel, a co-founder of Punta Cana, and his late wife, Ann S. Kheel.

Ted Kheel, along with the Dominican entrepreneur, Frank Rainieri, are the original owners and co-developers of Punta Cana, and the school is another example of their strong commitment to social and environmental responsibility. Education and helping the disadvantaged were Mrs. Kheel's lifelong interests. She served on numerous educational commissions in the United States and would have been proud to have a school named after her.

Located on the eastern coast of the Dominican Republic, the high school is already serving 125 students from the towns of Veron, Bavaro, Cortecito and Cabeza de Toro. With a total investment of $785,000, the school has four classrooms equipped with computers, as well as physics, chemistry and biology labs. There are also basketball and volleyball courts, administrative offices, and a cafeteria and restrooms. The second phase – to commence early in 2006 – will include the construction of a library, additional classrooms and workshops.

The opening ceremony was attended by Dominican President Dr. Leonel Fernández Reyna; Secretary of Education Alejandrina Germán; and Grupo Punta Cana founders Frank Rainieri and Ted Kheel as well as its directors, Oscar de la Renta, Julio Iglesias and Robert Kheel.

"The opening of this polytechnic school is one of the socially important projects being developed by Grupo Punta Cana," said Rainieri, "whereby poor children and teenagers from different communities in the area can get quality education."

Secretary of Education, German, emphasized the relevance of the Punta Cana initiative and "its outstanding contribution to the development of the Eastern region. It is my hope that this project can become a model to be followed in other provinces," she said.

the hell are you and how do you speak Spanish like that. By that time I had been in the country for almost two years and spoke like a peasant. My Spanish is good but I used terms that were very colloquial to the area. He said to me and I'll never forget it, "Instead of throwing rocks from the outside and standing on the outside with your arms crossed, why don't you work from the inside to change it." And I said, "Somebody has to give me a job." And he said, "Show up tomorrow at Aeroporto Herrera, which is the small airport in the capital, not the big one, and we'll fly out to Punta Cana." Mind you, I had been taking public buses everywhere for two and a half years in this country and all of a sudden I was getting on a private plane and flying with the CEO to the resort at Punta Cana. I had to go out and buy a dress because all my clothes had been beaten over rocks for the last two years.

At Punta Cana, Mr. Rainieri introduced me to some other people and took me around the property. It was 1992 and all that was here was the airport, Club Med, and the Punta Cana hotel. Veron, the community right next to this one had only five or six houses. Today it has 500 to 600. It was very, very isolated. Some people spent the day showing me around. They had just opened up the Nature Walk, a very pristine area, and had some idea of what they wanted to do. That evening in his office, Mr. Rainieri asked me, "So what do you think you could do?" We threw out some ideas. He very much wanted to create an environmental opportunity for visitors, the nature walk, and the interpretation. He was very interested in educating the employees about water conservation, energy conservation, and trash. Dominicans are very unconscious about waste. It's not like the States where you have a trash company that comes every other day and picks up your trash. You deal with your own trash here. Nobody picks it up. It just gets thrown outside. He also wanted me to work with the school, which at that time was a small open-air structure and probably had thirty to forty students. It had kindergarten to sixth grade then, if I remember it correctly. There were two teachers and they would teach simultaneously. On one side of the classroom, there was the kindergarten and first grade, and on the other side second and third grade. On the other side of the schoolhouse the other teacher would have the other grades. It was a very challenging task for the teachers.

A lot of the children wouldn't have gone to school if these facilities didn't exist. As a matter of fact, when they got to the sixth grade—where we cut off—they had to go to Higuey, which is 45 minutes each way by public bus, and they

One of the organic gardens which caters to some of the resort restaurants.

CONSERVING NATURE/BUILDING HUMAN CAPITAL

would have to pay for it. So they gave up their education and went to work, although you are supposed to go to school until you are 16.

There is a tremendous opportunity for the private sector throughout the Caribbean--but specifically in Punta Cana—nobody does what Punta Cana is doing. My department has a $1 million budget each year for the various activities. The government in these resource-poor countries does not put in the infrastructure that it is supposed to. When I came to Punta Cana ten years ago there were 3,000 hotel rooms in this zone. Today there are 30,000 hotel rooms with an eye to build 45,000 to 75,000 rooms. This area produces 35% of the GDP of this country through the airport and the other resources that are here. It is the biggest tourism destination that the Dominican Republic has. Nonetheless, you cannot get them to fill the holes in the road. You cannot get the government to pick up the trash. There is no high school. We are building the high school in the community next to us. We did a census of how many kids in the community go to high school in Higuey after they finish intermediate school. There are 400 of them. So we are building a high school there. It will have four classes, a computer lab and a biology lab.

We are also building a clinic next door. There are no medical services for people on this side of the island. So again we are working with the hotel association and Punta Cana is spearheading the effort until the government steps up to its responsibilities

So now we are at the point of why do we do what we do. And that's been one of the joys of working with Frank and Ted. Ted is very much focused on the environmental aspects. Frank is very focused on the social. He is Dominican.

Why does Punta Cana do it? Besides the fact that we need a school, it helps us to keep employees. This is an area that has zero percent unemployment. If there is zero percent unemployment and you are dissatisfied with your job, it is really easy to walk out of your job and find a new job the next day, especially if you have good skills.

Besides helping us to keep people it helps us to educate people. Among Don Geraldo's children, his daughter is going to college. That's the first person in the family who will go to college and she can do that because Punta Cana provided her with the ability where she could go to college. So we are training people who will come back and work for the company. And there are other people here, as well, young people who grew up here went to school here and came back. We need that. The DR has a very high rate of illiteracy so it behooves us to train people so that when they come into the work force they already have their basic skills. We consider that a need. The other thing that a school does is provide a community center. Apart from being an educational base for the kids, it offers adult-education classes. At night we do computer classes, language classes. You can get a master's degree. We are offering people an opportunity to continue their professional training and improve their skills.

When I first sat down with Mr. Rainieri, he asked "What can you do?" And we talked about four things: the two-room schoolhouse, the women's group, the ecological foundation, and teaching the employees about environmental stewardship. He asked me how much it would cost to do it, I gave him a figure. He doubled it and offered me a salary on top of that. And I said, "I'll have to think about it." I was very much dedicated to finishing my service in the Peace Corps and completing the school that we were building. We had a mutual friend, the director of an organization called Pro Natura, the biggest non-governmental organization in the country working in environmental affairs. So Mr. Rainieri called up this man, Jose M. Martinez, and said, "This crazy gringa has said she won't take this job." So my friend got in his jeep, drove four hours into the interior of the country, and sat me down and said if you don't take this job, I never want to hear criticism from you about tourism again. This is the opportunity of a lifetime. You've worked really hard for the school in the community. They have the money now and if they can't build it now then they don't deserve to have it." And he drove me an hour to the nearest phone so that I could call Mr. Rainieri up and take the job, which I did. And it's been a blessing.

When I first came out here, I'd joke with people and stand in front of the toilet and flush it, because I had no running water for so long. That and the dessert table. I'd been living out in the bush with no running water, no electricity,

Conserving Nature/Building Human Capital

mosquitoes all over and all of a sudden I have this job. They are paying me for what I love to do and I am staying in this five-star resort. I remember telling my mom I couldn't have a better job. And it was easy for me because I had been in the country for two years. When I did the employee education, I could speak to their reality.

When you talk to people about health and hygiene in the kitchen you've got to remember that most of the people don't have running water in the house. This whole concept of after you go to the bathroom you have to wash your hands before you go back into the kitchen is not just setting up a procedure You have to talk to them about why you are doing this and the difficulties they face and the realities of their home. I enjoy a very privileged status here because I am Do- minicanized in that sense and people enjoy working with me because I understand the circumstances they have.

When we started the environmental education, the trails, they were a great success. People loved to go through there. It was easy to do and we asked the university to come and help us. I am not a scientist and I'm not an environmentalist in that I am not trained as a technician, but what we have been able to do is bring in various resources that can feed into that process.

In the women's group, I decided to encourage handicrafts and decided to sell them in the hotel gift shops. We had to go around begging the visitors to buy the Christmas ornaments so that they would all sell out. I've never had a more empowering moment in my life than when I went to take the money that each of the women had earned, back to them. It changed dynamics. As soon as some of the women who had never worked before in their lives saw that they had earning power on their own, the dynamics in their home began to change, a change that was threatening to some of the men. I actually had some run-ins with some of them.

And Mr. Rainieri and his wife, Haydee—she's a powerhouse—went there and talked to everybody. He said, "Look what I've achieved. I've achieved because my wife is an equal partner in everything I do. You need to understand that your mates are also your equal partners in everything. You shouldn't be threatened by this. You should see this as an opportunity." After that things smoothed themselves out and I continued to work with the group.

Then we worked on developing the school. I did that for two years. It was the best job and it was one of the happiest times of my life because it is work where you feel that you are promoting a bigger agenda. You have one-on-one relationships with people. Now they call you because they have found a snake. But instead of chopping it up, they ask what they

Conserving Nature/Building Human Capital

should do with it. People have made that connection. This is what we do now.

As Punta Cana grew, one of the things we needed to do was to institutionalize the program within each of the businesses. It wasn't going to be the ecological foundation doing education, it needed to be the hotel doing education programs for their people. It needed to be the electric company, the airport. Everybody has to do their own program. Which means that if you detach these functions and make them part of the ecological foundation, then each of the businesses doesn't have ownership or a responsibility for what needs to be done. We created a corporate structure and that corporate structure has two NGOs that I run and they do their activities. But then I also have people who work in the corporate office whose job ii is to implement these programs with each of these businesses. I have ten managers and a hundred people who work, in addition to the ten managers that work in the environmental-affairs department. I can't think of any company that can say something as broad stroked as that. That is going back to the fact that resources allocated to ecology and conservation in Punta Cana are tremendous.

One of the assets we have is a vast amount of land. One of the things they have done is create three conservation areas. There's a 2,500-acre nature reserve that creates a corridor between us and our neighbors. That's the park.

We have the second-largest coral reef in the country that parallels our property here. It's declared a protected area against certain types of fishing. Now we are working on having it declared a national marine park. Hopefully, when that happens, the management will not be in the hands of the government but can be passed off to an NGO such as an aquatic foundation. At the very back of the property where the aquifers originate—a piece of property between CapCana and Punta Cana called Oyo Claro—we have had that declared a protected area as well. So we have buttresses on three sides of our property now that are conservation areas.

Jake Kheel

Jake began working with Ted Kheel developing models for sustainability through education, applied and basic research, and public awareness in 2001. He now serves as Environmental Director for Grupo Punta Cana and Director of the Punta Cana Ecological Foundation.

I STARTED WORKING IN PUNTA CANA in 2000 and worked for the Ecological Foundation and for Eloy Rodriguez. I was there for eight months and then went to Cornell for graduate studies. What I found in Punta Cana was that there were problems, but they were problems that needed to be solved and could be solved.

Grupo Punta Cana has made a major investment in advancing research and education programs concentrating on Dominican biological resources though the creation of the Punta Cana Center on Sustainability and Biodiversity. The Punta Cana Center is a state-of-the-art facility that conducts research, education, and conservation programs on the biological resources of the island with a growing list of international and national universities and organizations. The Center has ongoing projects with Harvard University, Columbia University, Cornell University, University of Miami, Virginia Tech, Stevens Institute of Technology, and Milliken University. The Punta Cana Center also collaborates with Dominican institutions, such as the National Aquarium, the National Botanical Garden, the Dominican Institute on Forestry and Agriculture, as well as the Museum of Natural History, and various

Dominican not-for-profit and community organizations. One of the primary goals of the center is to build the capacity of students and scientists in the Dominican Republic and help to bridge the gap of research and training programs between international and Dominican institutions. In this way, the Punta Cana Center has become a gateway for exploring the unique biodiversity of the Dominican Republic.

Punta Cana is interesting enough to get researchers to go there.

It has coral reefs, fresh-water lagoons. It is easy to get to, not like some of the other biodiversity areas which are almost inaccessible. Eloy Rodriguez gave a talk at Cornell about going to the Amazon, about the logistics and how complicated it is. It makes research very time consuming. Brian Farrell says Punta Cana is small enough to be doable, big enough to be interesting.

The creation of the Punta Cana Center was driven by Ted. Without Ted, there would be no lab. It has become a major point of interest for the universities. It also holds tremendous interest for tourists.

Applied research is very much what we are after in Punta Cana. It has potential for revenue as well because we can come up with a local solution that can solve local problems and can be applied elsewhere.

Sample coral from the reef is used for instruction in the lab.

Inside the plant nursery, hundreds of specicies are cultivated for landscaping around the resort.

Most beach tourism involves some labor, some jobs but the multipliers are simply not there. But at Punta Cana there is a lot of curiosity when people find there is a biodiversity lab. A lot of people don't go beyond the beach. But for those who are willing to take a walk through the woods, there's a lot for the researcher and the environmentalist. A lot of people want to travel to places and not just sit on the beach. For them, the laboratory makes the visit all the more enriching. There's golf, there's reef studies, there are the dunes and then there's environmental programs.

The University of Miami has made an obvious choice in selecting Punta Cana as the biodiversity model in the Caribbean it wants to study. Costa Rica and the Amazon have eco-tourism and backpackers. But the Caribbean model is researching the beach tourism.

The University of Miami is interested in the reef research in Punta Cana. We are dealing with almost 14 kilometers of reef, a fairly confined eco-system that is stressed for nutrients, overfishing and on-shore pollution. For Miami, it is a perfect model. It is accessible and it is in the Caribbean.

The Dominican Republic has now become a place where people show up. There are a number of interesting scientists. You can meet the key people in government. Everything is manageable. Ted's involvement means that he is able to attract top-level people to the place. He's developed an interested cadre of interesting people.

One of the reasons Punta Cana is unique is that it is owned by Dominicans. With Frank at the helm, they take great interest in ecology and the environment. That is one of the major virtues of Punta Cana that often goes unrecognized. And what Frank is doing is keeping in mind the entire community. There's a secondary school, a polytechnic. The multinationals that own resorts in the area are almost afraid to do anything in the DR because if they do something here they may also have

CONSERVING NATURE/BUILDING HUMAN CAPITAL

Butterfly catalog.

to do something in the other island communities where they have a presence. You look around and see many of the hotel and resort owners, most of them non-Dominican, who simply don't care how they impact the environment. Frank seems bent on serving the community and the country.

Brian Farrell

As curator of entomology at Harvard University's Museum of Comparative Zoology (MCZ), Brian Farrell succeeded the legendary sociobiologist E.O. Wilson. "The intellectual history of Farrell's approach can be traced back through three generations of students-become-professors to the late G. Evelyn Hutchinson, a biology professor at Yale. Hutchinson was an expert in the ecology of the niche. Farrell, by contrast, studies the evolution of the niche," says Jonathan Shaw in the Harvard magazine (Sept.-Oct. 2003).

Farrell began working in the Dominican Republic at the Jardin Botanico Nacional in Santo Domingo, one of the biggest arboreta in the world. He has expanded his research into working all over the Dominican Republic, including the biodiversity lab in Punta Cana where he takes students and lectures.

THE BIODIVERSITY PROJECT IS WORKING VERY WELL. We are in an unusual setting, in a developing country where biodiversity is among the least known. But given its proximity to the U.S., and its location in the Caribbean—one of the top biodiversity

A marvelous specimen from the rainforest.

hotspots—this is a very compelling research spot. Surprisingly only one-third of the insects in this area have been catalogued.

During his time, Trujillo improved crops but in the process he contaminated the countryside with pesticides. Since that time, very little active research has been done to understand and catalog the insect population that exists. During the same time, equivalent Cuban institutions did a great job documenting the insect population because they had a history of home-grown scientists who were trained abroad. Still, the Dominican Republic has one of the largest underserved insect populations and absolute biodiversity. It's a great jewel and, as yet, undiscovered.

There's a tremendous emphasis on sustainable development here. And the reason it works is due to the close relationship between business and industry. Costa Rica also has a very sophisticated research operation, but it is isolated in the hills. Here there are real synergies between research and industry. What we are learning, we can apply. The implications of our project for eco-tourism are immense. Indeed, the line between intellectual interest and aesthetic interest overlaps.

When Ted Kheel suggested the idea of a biodiversity lab several years ago, a project that would be funded by businesses here, I was not sure that it could work. Especially because we are so remote from the rest of the country, so dry. But combine the lab with a need for sustainable development and its proximity to hotels and resorts nearby and it is working. And it is self-sustaining, as Ted had expected.

CONSERVING NATURE/BUILDING HUMAN CAPITAL

Eloy Rodriguez

Dr. Eloy Rodriguez is currently the James. A. Perkins Endowed Chair at Cornell University and possibly the first U.S.- born Hispanic to hold an endowed position in the sciences at an American university. He is also Director of the Cornell Biodiversity Laboratory at Punta Cana and the Cornell Amazon Station and Laboratory in Peru. He is also the Associate Director of Drug Discovery and Herbal Pharmacology for the Weill Medical College Center for Complementary and Integrative Medicine (CIM) at Cornell University in New York City.

I AM A RESEARCH SCIENTIST and among the things I do is a lot of international research. I have another facility in the Amazon, a laboratory with an eco-tourism kind of lodge. I have always done research in Latin American countries, actually all over the world, but I also have been working in Africa and India. And I generally engage students in the process.

I am into drug discovery; I had some conservation issues in trying to argue about the resources that you can tap for medicines. I was trying to make an argument for conservationists. I developed a field called zoopharmacology, animals medicating themselves. I worked in Africa studying chimps and studying apes and discovered that they actually go out and seek certain things as cures. These animals are important resources in enabling us to better understand medicines. I was trying to argue that besides giving oxygen and fruit there is another component to the richness of the forest—medicines—and that animals in the forest maintain good health. In that context, I would take students to the whole world to interact with the local communities, the indigenous people, the mestizos. In India, in Africa, the same thing. We would live with the tribes and try to understand their view of the world, their sense of sustainability. In the Amazon, we went to study how they utilized Brazil nuts, how they got the oil from Brazil nuts, how they stopped logging and digging for coal when they saw a better return from having Brazil nuts and now they have control of it. In the end, it did involve tourism.

I met Ted when he came to listen to one of my lectures. I was talking about how I took students on eco-tours but that eco-tourism wasn't doing too well. So Punta Cana opened itself up to the idea of bringing in students who would study biodiversity, cataloging species, etc. What they wanted was information they could give to the tourists about the birds. They weren't thinking of conservation but later on they started to think of maintaining their biodiversity because it could provide some capital. I just talked about the experience of the students from Cornell. And Ted, being such a strong supporter of Cornell, proposed the idea of a biodiversity laboratory.

When I initially thought about the laboratory I wasn't thinking about eco-tourism, I wasn't thinking about conservation, I was thinking about what a great experience it would be for the students, a great experience to educate the local people about the importance of their resources. I was thinking of all this, not so much as a support of the tourism in the area but as a means of capacity building within that community. In other words, exposing people to the idea of better maintaining biodiversity because it is good for tourism. If the workers and the employees knew about it, it would be great. And if the tourists knew about it, it would be great too.

For tourists, it could be really important. After four or five days in a place they tend to get really bored. That's when we said that the biodiversity laboratory, the way I first thought about it, could be a nice scientific resource. We'll catalog all the species, we'll tell them what is rare and is poisonous, prepare ways that people can go and see them. But to do it right, you need some scientists.

We originally thought of the place as a place to train students—both Dominicans, which we do, and Cornell students in biodiversity and conservation and natural health/ecological health. From there, we began to look at things we could do for tourists. Then the question was: how can we generate some capital?

We wanted to look into conservation issues. But there was a dilemma. We couldn't push the conservation issue too strongly because that might be implied as being against development and we would be building. The issue was that if you going to be building golf courses, developing, you should know what you are destroying in the process. There are natural habitats that are destroyed in the building of golf courses. A golf course is the antichrist of biodiversity. I felt our role was to sensitize not only Punta Cana, but also all the other resorts about the impact they are having on biodiversity. When you do this kind of development you need to have people—not people that criticize you for development—but people who will make you more conscientious about what you maintain in your resort because that will be more attractive.

Conserving Nature/Building Human Capital

Guided walks through the nature reserve familiarize tourists with this poisonous tree.

Take the idea of rare species for example: Punta Cana has the second-smallest bird in the world. Bee size. The smallest one is in Cuba and it hasn't gotten a visa to leave.

These are the kind of things, I think, that captures the imagination of the people. They understand that it is nice to be in the beach and in the sun. They go there to enjoy themselves and not be annoyed. But in this unique area, the role that I thought we should play is to bring in some scientists to make the more sound decisions of what will be built.

But then I began to think more development, more about tourism. I thought about landscaping ecology. Things I had never thought of before. The coupling of tourism and the business world. We butted heads. There were times we said we should turn all this into a giant boutique. There was resistance. But it also permitted a broader discussion of the issues. The president would come. Government officials would show up. The biodiversity lab began to get more and more accepted as a unique thing to have within a resort. It was unique. It is not eco-tourism. That is different. Nonetheless, I began to realize the importance of the information we were generating and how important that was in helping make decisions. Cap Cana, for example, began bulldozing palm trees. Instead of that they should been thinking about ways to integrate them into the entire landscape.

When I began to look at the coral reefs, I told Ted I would bring in other institutions. So I brought in the University of Miami and Stevens Tech. The University of Miami is already working on coral-reef restoration on its own. That was a very smart move of Ted to help bring in scientists. Then I brought in other scientists, other specialists.

The most crucial thing in this whole Caribbean development is water. What are we doing about water? Instead of trying to simply be a research place, the biodiversity lab can become a major educational space. And they need to go looking into water quality. That would be unique, not just for Punta Cana but for the whole area. And I think they will do it. You see, water is the gold of this area and if people aren't careful they will screw it up. For example, by bringing in salt water to create a marina you can really distort the water balance. What began as a way to study the place has now evolved into something that is much larger and much broader and it has more of an impact than when we first began.

The engineers at first probably never thought about these issues and now they are because they have to. It's really the engineers, the construction people who have to have some kind of environmental consciousness. To understand the impact of what happens when you put salt water into a golf course. I am not in a position to advocate not doing something but rather to point out what can happen. You might have some incredible *cenotes* or waterholes that are ready to collapse and you don't want to build a golf course on top of that.

There's all kind of things that are important to understand. Now we are beginning to realize that when a golf course is too close to the water, the nutrients, the phosphates are really percolating down and the rivers take them down to the ocean and that's what is really causing the bleaching of the reef. It is this kind of understanding that makes it really important to have a research component.

I really think that this is a unique role for the facility. I also said to myself, "How bizarre to be playing this role." I didn't think that the tourists were going to go out of their way to see

CONSERVING
NATURE/BUILDING
HUMAN CAPITAL

a little biodiversity lab unless there was something there—like a little museum. I designed a whole museum with a garden complex. For the biodiversity lab, we were going to put in a medicinal garden of plants. We knew we were doing something different that tourists would like. The role of the biodiversity lab is to be a Disney World of the environment.

The other component is the educational—not just of the students but of the employees. There are nearly 2,000 people employed there and you have to provide something to the employees that has an educational value. The employees call this the university. They call it <i>la universidad.<i> We have Harvard, Berkeley, Cornell, and we have Dominicans all come to a center where the best institutions are. And now I am establishing a relationship with the Europeans.

The biodiversity lab has created a unique way of thinking about sustainable tourism. If you want to think about this properly you need it to be done by scientists, not by just ecotourist managers who don't have the tools to make the right decisions and the right judgment.

It also has generated this whole idea that it is good to have a center like this because it takes care of tourism and provides opportunities for training for the staff on issues that are very relevant about the environment, about pollution, about disease problems. Many of these ideas that are taught are good for the hotel. People love to come here and teach. It also has an impact on us scientists—a real hands-on experience. Now you know what sustainable development is and what the problems are.

I think it has attracted tourists, a lot of articles. People are very curious about it. I talk about capacity building; Punta Cana is a real-estate entity. The residents are going to be there and now we have a center that will provide a whole range of things for the new residents as well.

This protected set of Mangrove thrives on the beach at Punta Canna, only several meters from a popular swim area.

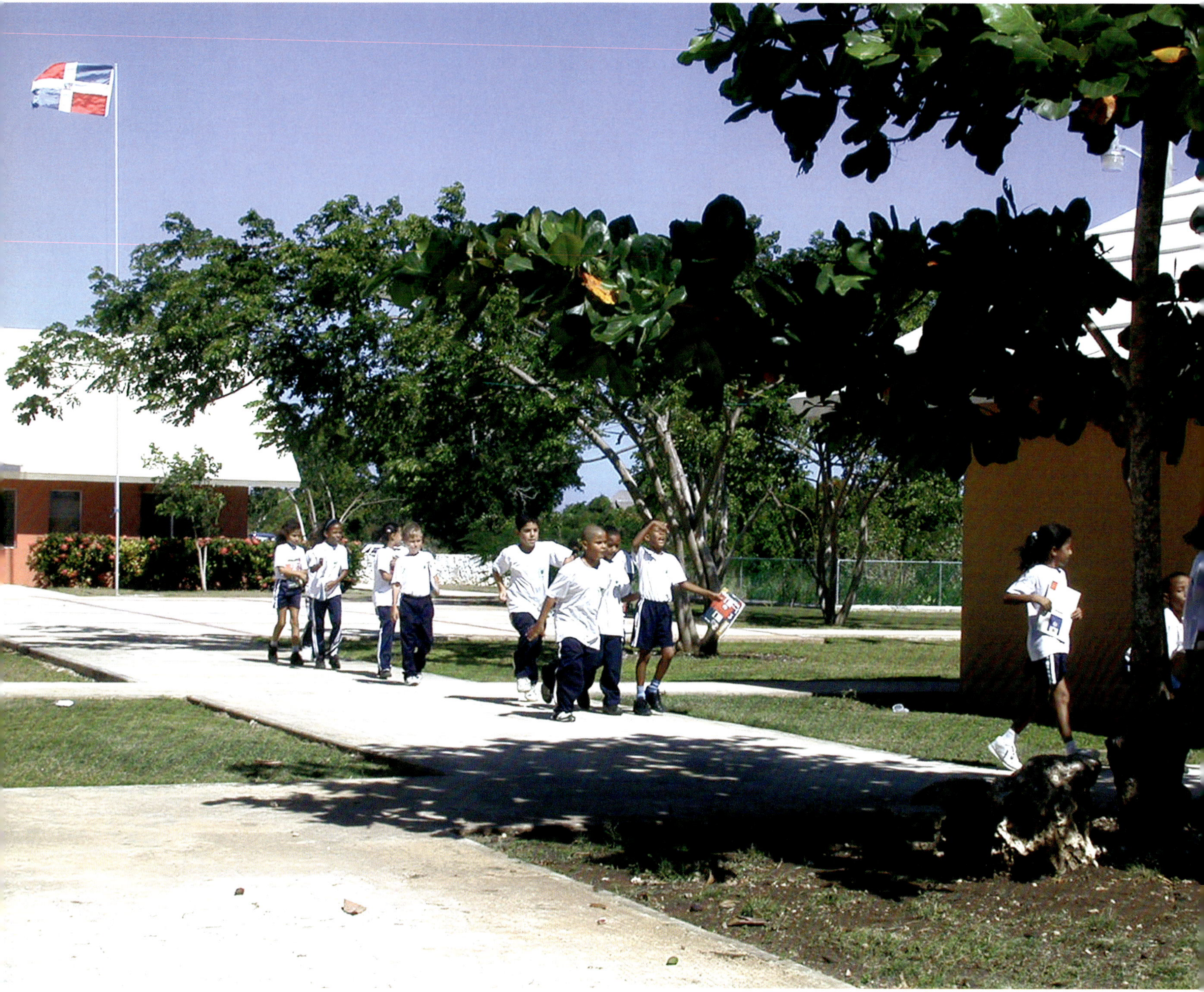

Human and Economic Development: Workers and Entrepreneurs Building Together

When stories of business successes are told, the focus is often on a few key characters, typically the founders. There is a tendency to do the same for Punta Cana. But Punta Cana, more than many other initiatives of its kind, is the collaborative effort of a slew of disparate forces. If the effort was limited to serving only the resort and its band of eclectic owners, the effort could very well have been still-born. Indeed, many tour operators in the region, mostly foreign, have chosen to do little for fear that they would have to do the same in other spots they manage.

Punta Cana recognized early on that its survival and well-being depended not only on its doing well, but also on the doing well by others.

A resort surrounded by poverty has only limited appeal, as does a resort that is difficult to access. But build the environs and you will help yourself. By contributing to the development of neighboring towns such as Higuey, Punta Cana has helped its own image. But there's still more to be accomplished. The economic priorities and incentives still need to be modified and changed.

Says Rainieri:

> We can improve the things locally in the resort area but we cannot improve the town. Still, the town has improved substantially. It's much cleaner now. The streets are paved but there's still no public bus service in the town. We've got 35,000 people who work in the hotels alone plus an additional 70,000 or so that are involved with the hotels. It's chaotic. The authorities haven't grown at the same pace as the industry because that's not their concern. They don't expect us to tell them what to do. They're concerned only with their own territory.
>
> There will be a big change in the next ten years. In this area the traditional support or solution was the landowner. But the important thing today is not to have land for cattle, but to have a hotel or tour operator, a restaurant or gift shop. That generates more income or creates more employment than having a cattle farm of

Construction in a new section of the town.

HUMAN AND ECONOMIC DEVELOPMENT: WORKERS AND ENTREPRENEURS BUILDING TOGETHER

2,000 acres for 10,000 head of cattle where you need only five underpaid guys to move them. So the dynamics are changing fast.

My perception is that in the next ten years at the most, there will be a tremendous change in policy and the town will become a 21st century town because of the presence of a new industry.

When we talk about the true success of Microsoft, it isn't the wealth that Bill Gates and Paul Allen and Steve Ballmer have amassed that is remarkable, it is the wealth that the company has created for thousands of its employees and for the many that are involved with the company. The same can be said about Punta Cana. Both Kheel and Rainieri have been successful with the dream that they persisted with for 35 years.

But it's the opportunities created for the more than 1,500 employees and for the dozens of entrepreneurs and businesses involved with Punta Cana that is the true barometer of their achievements.

The following narratives are from a cross section of Punta Cana—Haydee Rainieri, wife of Frank Rainieri; Gerardo Garcia, one of the company's oldest employees; Rafael de la Cruz, the company rancher and herdsman; Alejandra Tolentino, former employee and now owner of a company that provides services to homeowners and the resort; Ingrid Herrera of Guest Services. Together their stories describe the challenges and the achievements that have made what Punta Cana what it is today.

Haydee Rainieri

Haydee Kuret de Rainieri, the wife of Frank Rainieri, was born in Santo Domingo. After trying her hand as a physics teacher at the Pedro Henriqez Urena National University in Santo Domingo, she decided to become a public-relations professional and in 1987 joined the Punta Cana Resort and Club. At Punta Cana, she was appointed vice president of Marketing and Development of the resort. But Haydee Rainieri is more than just a spouse working alongside her husband, she is greatly responsibly for the tone and the philosophy that sets Punta Cana apart from all the other resorts in the area.

WHEN I GOT MARRIED I WAS not in this business. Then I changed my career. I realized I had to if I wanted to keep my marriage. In those days you were invited for dinner at eight and it went on for hours. So sometimes Frank started work in the morning, went to a business dinner, and we never saw each other. I told myself if I go on like this for several years there won't be a family or a marriage. So I changed my career to fit.

I worked first as a teacher, then in public relations. One day, because Frank couldn't pay a project manager I came on to do the work. I was always involved. I would go with the ladies for cleaning, checking this and that, and helping in the evening, greeting people, asking how everything was..

The board of directors assigned me a salary but Frank said: "No. You cannot accept it, it is too much money." I took the salary down. Years later, I wrote a letter saying I was quitting. "You are not paying me the salary you are paying others. And I am doing the same kind of work and have the same kinds of responsibility," I told Frank.

We began working together by circumstance, but then we began to really work together. And we have built all this together. Frank isn't your typical Latin American man that seeks to control everything.

Before I got married, I would go into the office in the afternoon to help Frank with things that needed to be done. After we got married, I would go into the office on weekends and I would meet the people in public relations, check the housekeeping and cleaning. The only thing I didn't get involved in was the kitchen. Finally, when we began building the hotel, he needed help and it was easy for me to work with him in a formal capacity.

Only once in my life did I say quit, but when I realized he wasn't going to, I focused on making it happen. And if the kids had to go with one pair of sneakers, that was it.

In this project we have been very privileged because we have found people that have taken on Punta Cana as a part of their lives. Our children were involved when they were kids. Now that they are adults and working for Punta Cana, I really think that they feel the same passion for Punta Cana that we felt and continue to feel.

I remember years ago our daughter Francesca asked, "Are we going to sell Punta Cana?" And I said, "I don't know. If it happens, it will happen." And she said, "How can we be in Punta Cana if it isn't our Punta Cana?" It was then that I began to realize that Punta Cana was as important for my family as it was for me and Frank.

When we got married he said to me, "You know I have a mistress and it is Punta Cana. You should join her." And that's what I did. I didn't fight Punta Cana; I joined it.

It was difficult in the beginning. Frank would go months without drawing a salary. If we couldn't get a salary, he needed to work on other businesses so we could live. So we had a disco, he distributed Bell helicopters, he distributed Beech aircrafts. I worked in the National Theater to make some money. There was only one time I told Frank to quit Punta Cana and leave because this gentleman offered him a job and it was a good job. We had three kids and we didn't have much money. So when he came and told me—we had been married over seven years—I told him, "We have three kids. It's not like before when we were young and by ourselves. I don't see an end to this. You don't collect your salary and we have to look around for everything. So I think it's time to quit Punta Cana and take a good job, a real job." And he looked at me and said, "Don't ever talk about leaving Punta Cana." And I didn't. It was the only time I said it, and thank God he didn't.

I began as project manager and with the construction of the hotel, I was dealing with the workers and the budget and the timing of the construction and all that. When the construction was finished, I began to deal with purchasing because there you have to deal with people you have to really trust.

HUMAN AND ECONOMIC DEVELOPMENT: WORKERS AND ENTREPRENEURS BUILDING TOGETHER

I began to get more involved. I went to operations, got involved with the hospitality area and then with sales. People often have problems with sales because salespeople are continually moving around. But they know that I won't be changed. I will be there. So they always trust that one of us will be in sales.

When we restructured the company I became vice president of Hospitality, Sales and Real Estate. Human Resources was an area we didn't know what to do with. We wanted to keep Human Resources officially under us. We used to have a director of Human Resources and when people had a problem they could go to him. But no one paid any attention to him. So we wanted to create a way that people would come to us officially. So I kept the Hospitality, Real Estate and Human Resources. But later, I gave up real estate. It went to my son.

But one way or another all the Rainieri family is involved—not in everything, not every day. But we try to stay in touch with people. Maybe there is one week I don't go to the hotel, or a month I don't go to the restaurants. It really is important to stay in touch with the people. Sometimes I don't know the names anymore. At one time I did. But now with more than 1,000 employees it is really hard. But we try.

In the business, Frank was the one who took on the social responsibility. I wasn't really involved with conservation or ecology. I learned that from Frank. I was more involved with people, helping them read and write, and teaching them cleaning and nutrition. But respect for the environment, that happened because of Frank. Little by little, I began to understand and respect as well.

I used to be the kind of person who would just glance at birds. Now, I enjoy them: their size, their color, their shape. I have learned to look at them through the eyes of nature.

When our children were younger we took them to the community to do volunteer work. What I have taught them is that it's important to share what you have, not just what you have extra.

I believed that our country is more advanced than many other countries in the Caribbean and that's because there are so many women in positions of authority and power. I think it's a matter of education. You can see that women are being educated even among the poorest people. Women are working and have economic independence.

Here, a woman can work, she can go to the beauty parlor, to the hotels or to classes on how to sew, how to cook. And it helps the men because they respect the women's independence. In the last five to ten years we have changed a lot, but we have a long ways to go. The men need to be educated too. They used to think of the women as their slaves. I used to say to our employees, during Christmas, when we were giving them the Christmas bonus: "This is money you were not expecting. Share it with your family. Don't spend it in the bar."

Education is really the key to development. Here we need education for everyone. We now have more women in our universities than men. We are talking about the family living and working together. And side by side with education is the notion of family. We go to Mass where there are many employees. It's important for them to see how we can all be together. That is why we try to fund many activities through the foundation. If people want to build a house, we tell them here is the opportunity, use it. If some of the parents don't go to meetings at school, I tell them to go. It takes time but they are learning and they are helping to teach others.

Alejandra Tolentino

Alejandra Tolentino got involved with Punta Cana by marriage when Manolo, her husband, started working in the Punta

Cana area in 1985. He was introduced to the Club Med by Mr. Rainieri. Grupo Punta Cana didn't exist at the time. When she finished school in 1993 she moved to Punta Cana with Manolo, to live and work there.

I STARTED TO WORK AS A DANCER. I had just finished school and I was just scouting around to see the possibilities. The possibilities were endless. You just had to take the chances. I started working with Grupo Punta Cana and at the same time Manolo and I started our own company. I suddenly came to realize that this wasn't just a job, it could become a lifestyle. I was working with Grupo Punta Cana to provide content for shows. Then little by little I began to work in charge of the stores here. They were being managed by Mrs. Rainieri and that was very interesting. She's very special, very strong, very focused and aggressive with people. She tells you exactly what she thinks. It was very interesting to see her and the Rainieri family, all of them, manage, integrate with Punta Cana, but each one do something different and still be part of Punta Cana.

When Dominicans speak about Punta Cana, they don't talk about Ted Kheel, Oscar de la Renta, or Julio Iglesias, they talk about Frank Rainieri because he's been the local figure in the spotlight for the good and the bad the whole time. We learned a lot from the Rainieris and we thank them for many reasons and for many lessons.

In the beginning, there was nothing here. People had to go to Santo Domingo and take a six-hour bus ride to come here. It was pretty tough. But people were young and it was an adventure. Manolo was introduced to tourism as a job—by Mr. Rainieri himself. In 1988, Mr. Rainieri gave him a call and said: "I am opening a hotel and I need someone to take care of the water sports and activities and entertainment." That's what Manolo was doing at the other places and he really liked it.

Manolo came here and started to put together the program. Punta Cana already had a mystique at the time and everybody who came here, came here just to sit on the beach and explore the natural walks, the lagoons. It was virgin territory and unexplored.

Then I realized that it was also an interesting place to work as part of this family because there was so much going on, so many plans, so much energy. You felt everything was possible. The way they talked about things to be done you were sure they would be done the very next day. They had the passion, the patience, the vision, the endurance to get here when there was nothing here. And when they told people they were going to build Corales, where there would be luxury homes and an airport that could handle more than one million travelers, people said they were crazy. They were going to be eaten alive by mosquitoes. They were going to be forgotten there. But now you go to the airport and see how much people depend upon the business it generates, how much everything has changed, how much everything has grown just because the airport is there—for the better—and you can't thank the family enough for being here. People like my husband and I who started as employees now have our own companies.

After working with the stores I was transferred to work with the resort. I had majored in social communications so eventually I got involved with the public relations of the hotel. It was important.

I had training in every department to get to know how everything worked. So when I started I had a vision of what was going on inside and outside. So, every meeting when we talked about clients, about procedures and dealing with people I could bring two perspectives to the table. I have been here as a "guest" for three years. I think I know what guests like and don't like. At the same time I know what the operation needs. It was interesting to sit down with people who

HUMAN AND ECONOMIC DEVELOPMENT: WORKERS AND ENTREPRENEURS BUILDING TOGETHER

know what public-relations messages to create and have their perspective, as well as the perspective from those looking in. I often ended up defending my clients. If I knew that some measures were not good for the resort because it would affect the way clients see the resort. I was the fighter. The bottom line for all of us was to create the right atmosphere for an unforgettable experience.

I manage properties now. We have a company that is divided into two branches. With one, we take care of the homes. The homes of Mr. de la Renta and Mr. Kheel; we manage those properties, among others. There are some cooks and butlers in those houses who earn more than professionals in the DR. They are service providers—like my business. Sometimes I sit at the table with the owners. We chat. We discuss financial decisions or improvements to their properties and sometimes I'm cleaning tables with the staff.

We own a company that focuses on services. It's the future for Punta Cana. As all these people build their homes they will need to rent the house if they don't have the money to support it closed for six to seven months of the year. They are going to need maintenance for their property. Real estate has a limit because you can't sell more than you have, but services are unlimited. Our company grows with the real-estate development. What we are doing can become a huge business, impossible to manage by just one company unless you want mass production and a very impersonal approach. But that's not Punta Cana. I manage properties that are already built and my husband builds houses. He has a partner in the construction business.

We are the present and the future of Punta Cana. We can take a person who buys a property all the way through from start to completion and management. We manage the Aurelio's property. We are very good friends and that's what we want to do with our business. We don't want a huge volume where you don't know who you are working with. That will be a need at a certain point, but we want to specialize in a certain quality of service. Instead of providing cheap service and getting many customers we are expensive because we are personalized. We deal directly with our clients, no intermediaries, and that's what you get after ten years of knowing all these people of knowing them from the time that a home in Punta Cana is just an idea in their head. I was one of the people who solved many of their problems at that time when they were guests in the hotel so the relationship developed. It was a logical crossover to have me take care of their homes. When I decided to leave the Punta Cana Group and start my own company Mr. and Mrs. Rainieri said, "We were expecting it."

I manage about 70 employees from very low educational levels to professionals. When people leave on their days off, then come back with a typical comment "There was no electricity, there was nothing to do. So hot."

There's a change in the work force from what it was 30 years ago to now. There was nothing here. Now there are 26 hotels, each doing business in its own way and when one of the chains decides to open a new hotel, they need new staff and experienced staff to start things moving. This staff will help them train others. They know that turnover is so high. It's tremendous. Some hotels have 70% turnover in a year. It's because there are so many job offers out there. People don't care. This hurts the results, the product. What do you need to keep qualified staff and certain standards? They need to feel it's worth it economically compared with what they can be offered elsewhere. You need to make people feel they can grow and move, maybe many steps, up the ladder. Not everyone is interested in that, but you want a critical mass of those who are. You don't need the employee who comes here just to get a check on the 15th and 30th of the month. You want the employee who will come, if you're offering English classes, and computer classes and German classes. The ones who are interested in doing everything. It's offered to employees and it's making a difference. With the school, new options are open. When teachers finish with children in the afternoon, they are working with adults, aerobics, English classes, computers, French, workshops on sales and marketing. These new options make people feel they're not just lost in the dark here—they have options—that's very important. With the situation in the country right now, an employee who goes out on his own and tests the job market can really appreciate the much higher quality of life that we can offer.

I still feel that I'm part of Punta Cana even though I have my own business. We use to say that this was the "Independent Republic of Punta Cana" and we had just one president.

Punta Cana is like a separate country. We don't suffer from electrical breakdowns or have the chaos of traffic. The state receives money not just from the airport, hotels, and business, but from people traveling out of here. Tourists come here and they go to Santo Domingo, Puerto Plata, and Samana, and money moves.

Milciades Guerrero

Milciades Guerrero began working at the hotel on the 11th of July, 1989. He had been working in the free zone in La Romana designing polo shirts when he heard that there were jobs available in the new hotel in Punta Cana. Through some contacts in the company's human resources department he managed to wrangle an interview and a job. He's been there ever since.

IT'S BEEN THE ONLY JOB I'VE HAD IN HOTEL WORK, but it's been very satisfying. When I started to work I had nothing. Here I've advanced substantially. I've learned English, a little German and Italian. I've been given the opportunity to learn different aspects of hotel areas of work. I've worked in reception, the restaurant, and telephones. The department I work in now represents the bell boys. I'm the captain.

I like this work because of the relationship with the clients and I'd like to stay in the maleteros department. It's the relationship with the clients that I like. It interests me to make sure that when a client arrives their needs are met. When clients arrive I usually inquire about how they are. When I take them to their rooms I ask them if they are satisfied with the room and also when they leave if they liked the room. When I can help them with anything, this is what I like—to help. Client satisfaction is my principal objective. This is also the principal goal of the hotel.

I haven't had the experience of working in other hotels. I've always been here. But many people who have left here to go work elsewhere, always say they want to come back here. Clients who make return visits always look for me and ask for me and I ask if there is anything I can do for them. This is a pleasure for me.

HUMAN AND ECONOMIC
DEVELOPMENT:
WORKERS AND ENTREPRENEURS
BUILDING TOGETHER

I'm very thankful to the hotel for giving the employees the opportunity to learn the entire business. An employee, for example, can begin as a bell boy, then work as a cook, then learn all the other departments to advance and be trained in all areas. Employees don't have similar learning opportunities in other hotels

I come from a pretty humble family, second of ten children. Three of my brothers work here and now my wife works here. I have a lovely wife and three children. A house and a car. It's not new, but everything I have I owe to Punta Cana. When I started working here I didn't have my own family yet. I met my wife here and then had our children. My children are 12, 8 and 7. We live in Higuey and my children go to school there. Consequently, I consider my family to be a part of here. I consider the hotel to be my home. In 15 years, I've been only out five times and been sick once. I've been recognized for my work, won awards, one was for "Leading Employee of the Year."

It's important that the company gives an opportunity to members of the entire family to work here. There are now eight people from my family working here. For many companies, only one person from a family can work. For the Punta Cana group, the company doesn't want to break the connection with the family. My ambition for my children is to have them get an education, to learn English and computers so that they can work as educated professionals and have a career.

I've had the experience of meeting well-known and famous visitors. It's been great to meet people such as Julio Iglesias, Oscar de la Renta, Shakira, and President Fernandez, but what I like best is that have helped many people to be satisfied with their experience in Punta Cana.

The most important changes in the 15 years I have been here?

The hotel has expanded and grown. The employees have grown to be part of a full-fledged company. I have been able to get my own house with the company. The number of buildings and the size and scope of the resort has increased. But the biggest challenge was the suffering caused by Hurricane George. We all had to work together to recover from that.

Gerardo Garcia

Don Geraldo is a much respected person in the Punta Cana community. One of the oldest employees of the company and one of its senior most, he has emerged as a natural leader and an "elder" spokesman for the community and for other employees. Every member of Geraldo's family has at one time or another been employed by Grupo Punta Cana. His wife is a housekeeper in one of the suites.

WHAT WERE THE PEOPLE LIKE, the early people who were here? The people, a lot of them aren't here but they all got paid their due. A lot of people were actually taught by Dona Haydee and Don Frank personally, which is something very important to be told. A lot of people have been educated, lawyers, school teachers, university students because of the training that was given to them. From the beginning, people started to come here, to carry on the work that Don Frank and Dona Haydee started and it has never stopped.

I was recruited in 1969. I have been here 35 years, from the very beginning. I started off as the chef here; I also was in charge of personnel.

In all the years that I have been here I have seen many changes. Before, the trip from here to Higuey used to take five to six hours. Now it takes half an hour to forty-five minutes. In the beginning, every Friday I used to go the airport for the runway cleanup, to start to fix up the runway for the following week. Today, the airport brings in thousands of tourists. It is a world by itself.

This area would be a very neglected area without the resort. It is an area where modernization wouldn't have come so quickly without the resorts, and people would be doing more or less what they were doing before. What the hotels have done is give the people here a new way of life.

Development has brought electricity and running water. And with the plaza, the school, the growth of the community here is very noticeable. And with that, there has been an interest in the environment. Ninety percent of the people are familiarized with the need to take care of the environment. Because there is a lack of resources, people realize early on the need to save what they have.

People are constantly urged to conserve energy to protect the environment.

I am very proud of where I live. I have my own plot to cultivate and I am also able to teach others how to do the same.

Without the resort, my life would have been completely different—90% less opportunity. We would probably have had to live in the city, totally cramped and with no room to bring up my children with any dignity. Here, I can control the environment and decide on the lifestyle and not be forced into it.

My life here, even though many will see us as poor people, is rich. The faith that is placed in me as a worker makes me very proud. We feel protected, we are protected, and that is hard to put a price on.

Rafael Andrés de la Cruz Harache

Born Feb. 14, 1938, Rafael spent his childhood in La Romana until the age of 12. When his father died, he was sent to live with her mother's family in the countryside in El Seibo, where he stayed until the age of 14. During this period, he worked on a farm.

Before reaching 30, he went to live in Santo Domingo with a relative. He became a policeman and after ten years of service, retired due to a traffic accident. Having nothing to do any more in the city, he returned to his family in El Seibo. There he married Digna, his first wife, with whom he had five children, two of whom died in traffic accidents.

When Rafael was in El Seibo, the family of Gerardo Garcia (the cook of the hotel and his neighbor), asked him to go to Punta Cana to inform Gerardo that his father was dead. He brought back Gerardo and after the funeral, they all went back to Punta Cana. There Gerardo encouraged Rafael to work at Punta Cana and spoke to Frank Rainieri to offer him a job. Under a coconut tree, Frank Rainieri interviewed Rafael and hired him.

In the beginning, Don Rafael worked as an assistant at horse service or country guardian. After three years, he took charge of the property overall for the security of lands. He had five men working for him to protect the property, mostly against people entering the land without authorization and cutting down trees to make charcoal.

BEFORE WORKING ON THE FARM, I worked for three years taking care of the plots in Punta Cana. Before that I was a policeman in Santo Domingo. In my third year Senor Rainieri put me in charge of the livestock and property. There are now 60 horses, 3,000 cows. We don't eat them here. We're growing the herd. When we have enough cows, we will start to use them. The horses are for the guests to use and for us in the maintenance of the property.

The idea of the farm really came from an idea to profit from this huge land and to do business with cattle-rearing at a low cost because cattle can be bred alone. The cattle help conservation, limit people from invading lands, and as we have to always visit those places where the animals are, it indirectly helps the supervision of the property. Any abnormal situation is immediately reported to the security.

Because the cattle are scattered in various places of the property, I do a daily supervision and verify that everything is under control and nothing is short. The workers in charge of different places where the cattle are kept are also keeping track as well.

The cattle are distributed in two parts of the plot, a fence from the Juanillo road and the other in the plots in the back of the airport. With the construction of the new golf course, the two groups will gather together. We also have goats and sheep in the place we call the 25. In each one of these areas, we have a man taking care of the animals. All of the employees at the very beginning of their job were coached in the daily routine, the special cares. They are daily supervised by me and my son, who was trained by me.

The nature of the animal care has changed over time. Currently, we have a veterinarian from Higuey who visits us once a week and helps us to prevent sickness in those animals. He also administers a preventive program against ticks.

When am I going to retire? I do not think of retirement; I really wish to work until the last moment of my life if God allows me so. I have Punta Cana in my heart and I do not know or think anything else. If God decides otherwise, I will stay at home enjoying my little children. My whole family has grown with the company and hopes to continue to.

If there was no company what would have happened to me? My two youngest children are in school. The oldest son is an engineer. Punta Cana has changed our lives. When you get to a place that offers you the kind of opportunities that Punta Cana offers, it is very difficult to leave.

Ramón Virgilio Rodríguez Pérez

Ramon Rodriguez has been working as a fireman for 14 years for the Punta Cana Group. Before that, he worked as fireman in Higüey, province of Altagracia. When he started in 1991 in the fire department, the airport was very small—between 5,000 feet to 7,000 feet. Since that time, the airport has grown a lot and developed. And air travel itself has become the subject of greater scrutiny and cautiousness. Airports themselves are seen as more vulnerable and in need of security.

I THINK I DECIDED TO BE A FIREMAN because I felt I had some vocation of service.

After the tragedy of 9/11, to be a fireman has taken on added significance especially because of terrorism and terrorist threats. But we do not feel scared as firemen. Still, I am aware of the risk, that there is a higher risk than before. The part of my job that I like the most is to save lives, when I can help to do so.

I live in Higuey but I didn't have any training to work in an airport. Here, there's no specialized training in how to work as a firefighter for an airport. In 1994, I began to get specialized training in fighting airport fires. The last three years have been the most intensive training with new equipment.

We came out of structural firefighting but hadn't trained in airport fires with fuel and materials in an aircraft. In 1991, we had structural capabilities like most firefighters. The last four years have accelerated the firefighting process, more equipment, more trainers, potential exposure from toxic materials.

When I started 13 years ago, of course, I imagined I might fight a fire on a plane but I didn't really know what the reality would be like. 9/11 had a big impact on us because of the numbers of people who died and the numbers of firefighters. When you work in the same field you can imagine what it would be like to be in the situation. But when you work as a firefighter you are working in a tradition of service. You know it's something noble that you are doing.

I think that this will be my work for my whole life, because when you are forming part of the fire team, it is impossible to leave it because you feel satisfied by helping others in emergencies. When I finish my duties as fireman in the airport, during my free time, I study and devote time in the fireman station in the city of Higüey. I will retire at the age established by norms.

Punta Cana, it's my house, it's my father, it's my family. When I began here I was married with one son. Today I have three. I was studying in intermediate. My condition has gotten better. Today, I have a proper house, then I didn't have one. My son is on the Yankees farm team. Maybe soon he'll move from the minor leagues to the majors.

My family feels proud of the work I perform, and the fact that I am a fireman gives them more security.

Ingrid Herrera

When Colombia-born Ingrid Herrera was looking for work, her university in Colombia sent her resume to various businesses and organizations in Colombia and other Caribbean countries. At that time, Alejandra Tolentino was running the public-relations department at Punta Cana. When a copy of the resume crossed her desk, she called Ingrid to ask her if she wanted to come to Punta Cana for an internship as a public-relations assistant. Ingrid had already finished the requirements for her degree and was working in Colombia. But she said, "Why not?" And she ended up in Punta Cana.

I HAD PLANNED TO STAY HERE a maximum of six months. It wasn't in my plans to live in the Dominican Republic. I wanted to be in Colombia close to my family. I also had a ticket for three months. But my plans changed after I came here. Alessandra became pregnant and had her baby, then went on

HUMAN AND ECONOMIC DEVELOPMENT: WORKERS AND ENTREPRENEURS BUILDING TOGETHER

leave. And I didn't want to leave them without an experienced person in the job. I said I cannot leave the job and go back and leave all this to someone who is inexperienced and irresponsible. That was seven years ago.

When I came here, the hotel was being constructed. At Corales, the only house belonged to Mr. Iglesias. The marina was complete, but wasn't really residential yet. Since then there has been a huge wave of building. There is a new golf course under construction. Then there are other projects like the church, the schools, and the social services

There are many opportunities to do business here. There aren't that many cultural differences between Colombia and the Dominican Republic. Santa Marta is a city next to the sea and things there aren't any different. The temperature is the same, the same kind of beaches.

I came here from a big country that is a little bit dangerous. In Bogotá, a woman alone is always afraid something will happen. If you want to call a taxi, you need to know exactly where the taxi station is and where you want to go.

Things here are done much quicker than when I was in Colombia. There are no excuses for why things cannot be done.

I have learned a lot. In Colombia I wouldn't have advanced the way I have here. Here I began as a public-relations assistant and then when the company reorganized—into public relations and guest relations—I moved into guest relations. The relationship/interaction between the employer and the customer in this area is different. Most of the owners in the area are international chains, they're not Dominican. They earn money but they don't spend the money to protect the environment or to preserve what is here.

Nicole Ramirez

Nicole was an accountant, a bookkeeper at the resort and advanced to the position of controller general. But she finally left because of the long hours and the children. Her husband, in the meanwhile, was promising her that he would find something that they could do on our own. They finally developed a business called Casa Holanda with Nicole's sister, who still works with Grupo Punta Cana.

I FIRST WORKED IN PUERTO PLATA in the hotels near the city but then came to Punta Cana in 1992. It was a very big change coming here because in Puerto Plata the hotels are nearer the city. Here it was the middle of nowhere. I came with a lot of illusions. There was actually nothing here.

One of the things that surprised me when I came here—although this was a big change, being so isolated—was the human factor. There was a lot of support and human warmth. I felt important to the institution, like I was part of the family. This was the reward when there was nothing else.

What created the feeling? Don Frank and Dona Haydee. They were very affable, very caring with a big sense of what family is. They made you feel it. This is one of the triumphs of Grupo Punta Cana, how important human worth is to them.

Casa Holanda is not really a gift shop—there are many in the area. The idea was that the area was growing and people would actually begin to live here and start to decorate their own home. Not only the visitors, but the people who work in the hotels who would want to have a home here instead of traveling into Higuey. We provide a service for the people who are going to live here. The store has a combination of things. We have typical furniture from here and rustic furniture from Mexico.

The garden center started a year ago. When I left Punta

Cana, I had been used to a very active life. At the store, things were very quiet. I was just waiting for customers. I had too much energy for this. I had to do something to be more occupied, to give me a sense of identity and satisfaction.

My husband is Dutch and we took a trip to Holland with the children. While we were there, we went to all of the big garden centers. I always had a thing for plants and flowers. I thought, we have so much space, why not add plants to the shop? When I said this to my sister and my friend, they said it was crazy. "You don't know anything about plants," they said. But you just have to know what you want to do and then do it right. I had a head full of ideas and decided that when I come back this is what I want to do.

So I called my brother and asked him if he could help me build a structure to grow the plants. We didn't have that much money because we had just taken this trip so it was a very small and modest beginning. We were able to get credit from the company when we bought the plants (30,000 pesos) and we had 10,000 pesos saved up. This was all the starting capital.

We had a very good reaction from the public. The first customers were word of mouth—from local people. Then the hotels came aboard and then the local businesses. Later came the part when people contracted with me to design their gardens. Though I didn't have a clue about how to do it, I went with my guts about how I would like it and the clients were satisfied as well. That's how everything evolved.

Even now in the world of plants and gardens I still get questions from people "What do you know about this?" I've never studied this. It's something I do out of instinct and feeling. I get ideas from where I buy, from other designers, and live with plants day to day and have acquired a lot of knowledge.

I have a very good relationship now with the Punta Cana group. I provide the orchid arrangements for the hotel lobby and the airport and the corporate office. Also I do other hotels' reception areas. One of the biggest tour operators here is constructing a new office building and I'm doing the main entrance. It's a substantial business now, not just a small garden center anymore. It's 20-foot palm trees and everything.

When I started, the income was 1,000 pesos a day. During the first three years the concept was new for this area. The infrastructure wasn't there yet. It was growing a little bit, but we were re-inventing the business itself so there weren't any big projects. But the area is growing. There's more of an infrastructure and we ourselves have more of an inventory. We also have more experience now. As a matter of fact we want to expand, to have more greenhouses. We need more exhibition space. Now the business brings in 20,000 pesos a day. The size of a project is bigger—300,000 to 400,000 pesos.

I've done three complete gardens now for private homes. If private homeowners need orchids or special plants they come to us. And since we have a combination of things other than plants and flowers, people also come to us for furniture or small gates or something else for their home. A lot of people at Punta Cana now know us and come here to buy. The garden business is plants and orchids and pots. The furniture business is separate.

When I first came here, I thought I would be here for a year or two at the most. But I got caught up in the development of the area and the expectations that were created. I felt that if I was going to do anything I would rather do it here where there were more possibilities.

Punta Cana now is one of the biggest factors in the economic development of the country. There's still a lot of land and room for development. There are a lot of people moving in from other parts of the country because we have electricity, we have water. In the beginning, a lot of workers lived within the hotels because there wasn't anything else or people were moving back and forth from Higuey. Now, a lot of people are attracted to the area like to a magnet. They are starting to get small houses. Before, someone would be working at a hotel and the family was wherever and they would work for three weeks and then see their family for three days. Now, they bring the family and stay in the area. It's a steady flow of people coming into the area. You can see it within Grupo Punta Cana. In the beginning, when we were married we lived in the hotel because there wasn't anything else. Now, just in front of the airport there's a housing development. So the trend is that everyone is starting to live outside the hotel and within the community and you get a sense of the "City of Punta Cana." It's not fully formed yet, but it's one of the things that visitors ask: "Oh! This is Punta Cana. Where is the city?" There's no city of Punta Cana, not yet. People expect to see it but there's just Higuey. No doubt in a few years you will have a city of Punta Cana, because there are little villages forming now and somehow they'll all connect.

Fernando Soler & Eladia Torres

When Oscar de la Renta moved to Punta Cana from La Romana (Fernando Soler was then the tennis pro at Casa de Campo) he told Fernando about it. Soler was about finish his studies in hotel management and was looking for a job in that field. Oscar said, "Come to my hotel here in Punta Cana instead of staying in Casa de Campo."

Eladia Torres made the decision to come to Punta Cana for two important reasons. The first was that Grupo Punta Cana as a company was very solid. The second and the most important reason was for her family. She was divorced with three children. And Punta Cana seemed a most agreeable place to live with children, quiet and secure and a school close to the house.

The two together now own and operate a salon and clothing boutique in the plaza, financed in part by Oscar de la Renta.

Fernando Soler

I MET MR. DE LA RENTA more than 20 years ago. I started in Casa de Campo picking up tennis balls for him. I was the ball boy. I eventually started to teach tennis at summer camp. Then, he asked me to come to Connecticut to teach tennis to his family and play with him.

But after Hurricane George there was nothing to do here. We were working very hard to get the resort ready for a very important visit from the Clintons. They were going to take about 100 rooms. It was a total mess. I started working with the architect and the engineer. When that was done, I joined the hotel force, starting from the bottom. I was a receptionist. Then, I went to the controller's office. Then, to guest services, some training in the kitchen, and ended up as manager on duty for the resort. But just before I took that position I left the company and went to the States because I was offered a very good job at a club in Connecticut.

The club in Connecticut was going to bring over my whole family—wife and kids—over a two-year period. But I wasn't feeling very comfortable with the weather. I was very cold and I couldn't put up with it. I told them I was sorry but I had to go back. After leaving and coming back, Grupo Punta Cana was a whole new face for me. So I started as manager on duty then I got an offer to be the room manager. I was doing that for a few years until I met Mike Davison who got me into real estate.

I was playing tennis with Mr. Rainieri, who asked me how I

felt about going into sales. "I think it might be very good for you," he said. And Mike Davison said the same thing, "I want you in the sales force." That's my Punta Cana employment history.

Eladia Torres

PUNTA CANA DEPENDS NOT JUST ON ONE PERSON but on four very strong people. It's a family business. If something happens to one person, another will take over. It helps that we are working directly with one of the partners who is very happy with the company and has invested in the company.

In many other companies they wouldn't be happy to see their employees doing anything else, but with this business we are adding value to Grupo Punta Cana in another way. There was nothing when we opened the businesses in the plaza and there weren't many people interested in doing anything there. Now, after we started, there are many more and nobody wants to get out. Everybody wants to get in but there is no spot anymore.

I think Grupo Punta Cana is happy what we are doing. It adds value to the plaza. Besides, they didn't have to put any money in. The store owners financed it themselves.

Fernando Soler

My wife's idea was to start a beauty salon in the area. She herself had to go about 45 minutes to get her hair done and drive the kids. She thought, as long as they were going to build a plaza, why not start with a beauty salon. Mr. Rainieri thought it was a great idea and we could be very successful. I told Mr. de la Renta what we were doing and asked if he wished to be involved. He said, "Well, Fernando, I have to be honest about it. Whenever I've been involved in a business I don't know I haven't done well at it. I don't know if this will be a good idea because I don't know much about it and I know you don't know much about it either." I said, "Yes you're right, but we feel there's a real need for the idea and if we do it well, and we hire the right persons we'd do well."

We started doing it by ourselves since he didn't want to be involved. We actually got a loan from the bank. One day when I was playing tennis with him he asked me if I had gotten a loan from the bank and I told him I got it and he said, "How much are they charging you," and I said 35% and he said, "What, are you crazy? You're not going to take that. Put it back. I'll put the money up."

Then he said, "I want to go with you to the spot where you are going to put the salon." We had picked a spot already. I was looking at it more as a financial issue. The spot we had chosen was nice and didn't cost much. The ones that were better located were much too expensive and we didn't want to pay more. We had already started laying the floor and I had talked to the builder about how it was going to be. But Mr. de la Renta took one look at the place and said, "We're not taking that one. It's too hidden. You need a space up front, so that the people who are coming have easy access."

I said to Mr. de la Renta, "That's a lot of money for what we are thinking of." He said, "No. No. No. We are taking this one. Not that hidden one over there." So we took the new one. My wife had an idea to take the new spot as something for the future. When the new spot looked too big, Mr. de la Renta said, "Let's put a wall and put something next door. We can put some of my clothes." All of a sudden we had the store. It's not doing great, but it's doing OK and it pays the expenses. We are thinking of it as an investment for the future. In a couple of years when this is all built up and there are plenty of people walking around it will be a great investment for then.

They were putting up a second phase of the airport, which is a closed air- conditioned terminal. We came up with the idea of putting up another store in there. We spoke to Mr. de la Renta and he said, "I've already been thinking about it. It's a good idea. Let's do it." So now we have two Oscar de la Renta stores along with the beauty parlor at the plaza.

Punta Cana is now only 25% developed and we are covering expenses. So as this grows, I think the public, just from Punta Cana, will increase sales much more. We are also counting on the whole area. There's a new highway that will connect this side of PC with Bavaro. Now it's half an hour. Then it will be ten minutes. The community will be much more united, not spread out and this will make a big change.

Another change will be finishing the plaza. They may put together plaza tours with clients and guests. There are plaza stores not open yet. A gallery, a supermarket, a commercial center. When the things the tourists are looking for are ready, many more will come. They will go directly to our store or they will go to other places and see our store.

My work in real estate is fun. We get a lot of pressure but also the satisfaction of having people believe in us and in the product we are offering and I look at it as a fun way of making money. You get to meet a lot of people and open a lot of doors. We are about to open a second golf course section of lots and home sites.

It's a nice way to make a living. I think this going to be my future. I have set goals for myself. I think I'll sell real estate for another five to seven years. Then I'll probably open my own business.

Grupo Punta Cana gives those who work for it a lot of opportunity. You just have to be there when they are available. I am not going to say that there are programs that give you this or that in any certain way. But they do offer opportunities. When they show up, you have the chance to go for them.

Our business plan is to succeed by providing the best service in the area and the best products available. We have the same vision as Grupo Punta Cana, top-of-the-line staff marketing to the top-of-the-line people. We publicize by radio commercials, fliers and the resorts and guest services. But word of mouth works best. We have a total of 15 employees between the airport and the plaza. Almost all came from Santo Domingo. We had to get people to believe in us like we believe in the company. Things aren't so great at Santo Domingo. Here you get a good salary plus commissions for what you do. Unlike Santo Domingo, there aren't a lot of ways to waste your money. You work hard and save. So after a while you can put money aside to do your own business here.

In the DR right now if somebody offers you a job in PC you have to think it over very seriously because PC is the most important part of the republic. People like to come here because they know that there are so many opportunities to improve their lives.

Mama Luisa

Fate brought Luisa Patrovita to Punta Cana in the late 1990s. She had owned a restaurant in Italy, on an island for many years. After she moved to the Dominican Republic (her former husband was Dominican), Frank Rainieri said to her daughter Elena, who was the chief of banquets at the Camera Americana that he needed a person to run a restaurant at the marina. He was worried about the lack of development of the marina.

Luisa (better known as Mama Luisa) visited the resort three times and had her doubts. Should she stay? Should she not stay? Should she do it? Should she not do it? Finally she agreed. "But you'll have to give me carte blanche," she said.

I AM DIVORCED WITH FOUR ADULT CHILDREN. The children are all here. I have two restaurants, one La Yola, the other at the plaza. I have been here seven years running La Yola with much love. People love me and because of that they call me Mama Luisa. My second restaurant in the plaza is called Mama Luisa.

This resort is completely different from the others. Grupo Punta Cana is more human and personal than the others.

Senor Rainieri is a constant point of contact for us. What we are trying to achieve is a cooperative goal. If there is a problem, we try to solve it all together instead of trying to assign blame. Mr. Rainieri is always looking for a solution from among us. We are a big family as Mr. Rainieri likes to point out. I think we are in a very warm place. We are working in an enchanted environment where there is no strife.

Thirty years ago everyone thought Don Frank was crazy. I knew Don Frank when he came here in a little jeep, on a horrible road and a little house on the beach. No one thought that Don Frank would be able to make it happen, not just think it could happen. But his personality and his drive encouraged people who worked here to work with him and fight for his dream

The La Yola menu is developed from local products—fish and shellfish. Naturally I also have Italian products—pasta, tomatoes, parmesan cheese that I import. I cook only in extra virgin olive oil. We don't have anything fried. It's a very Mediterranean cuisine, with dishes such as a spaghetti frutta di mare with shrimps, calamari, langoustines. The food at La Yola is very healthy. I educated the personnel when I came. Today, I have a head chef who was a steward at another hotel; the bartenders also were stewards at another hotel. For many years it's the same group. They take care of everything at La Yola because its like something of their own. It's truly a big family. I love that it's a simple place where we work together. Sure there are problems, but these are everywhere. They are problems we can address and look for ways to resolve.

I've had a very nice clientele at La Yola—Bill Clinton, Carolina Herrera, Oscar de la Renta—all of them are there in pictures with Mama Luisa. Our regular customers are the hotel managers, the people who go to Club Med, the boat owners. There is a fishing tournament, there are about 50 boats fishing for marlin. The marina is a very charming place, very pretty. I live in the marina in an apartment. It really is an enchanting place.

When this hotel is slow I am protected because I have guests from the other hotels—also the bar owners. Guests come by car, there's a ferry boat that comes from Club Med. Always there are groups. I prefer to serve people by the plate, not in a buffet. Sometimes the fishermen are unloading lobster from the boats and people see it and say: "I want that."

I have a good relationship with local suppliers. Not a lot of paperwork. I pay cash. A fisherman can't wait for a check. I have two to three fishermen I work with, As for the Italian products, I can get everything I want directly from Santo Domingo

La Yola belongs to the resort and I manage it. Mama Luisa at the plaza is mine and my niece Rosanna manages it. It's a typical Italian trattoria, open 10 months.

How did one thing lead to the other? Rosanna was living in Italy and was married. But after she divorced she came here/ I thought, "I am going to give her something that will make her feel better." She is not only happy and committed to this, she is not thinking about anything else.

The restaurant at the plaza is Mr. Rainieri's idea. The plaza began with the bowling alley, then the pharmacy, then the bank. Again, people said it was impossible. Now there's a Dominican restaurant, a Portuguese restaurant, an Italian restaurant a bar, a nightclub. It's all coming along. There's a shuttle every hour from the resort to here and back.

I had a property on the island thatI sold to finance the restaurant in the plaza. The sale covered all the costs of opening here. Everything began without doubts or worries and it's growing every day. I'm beginning to recover everything I spent and I am beginning to feel very good about it.

For me, I didn't look for the Dominican Republic, it came to me.

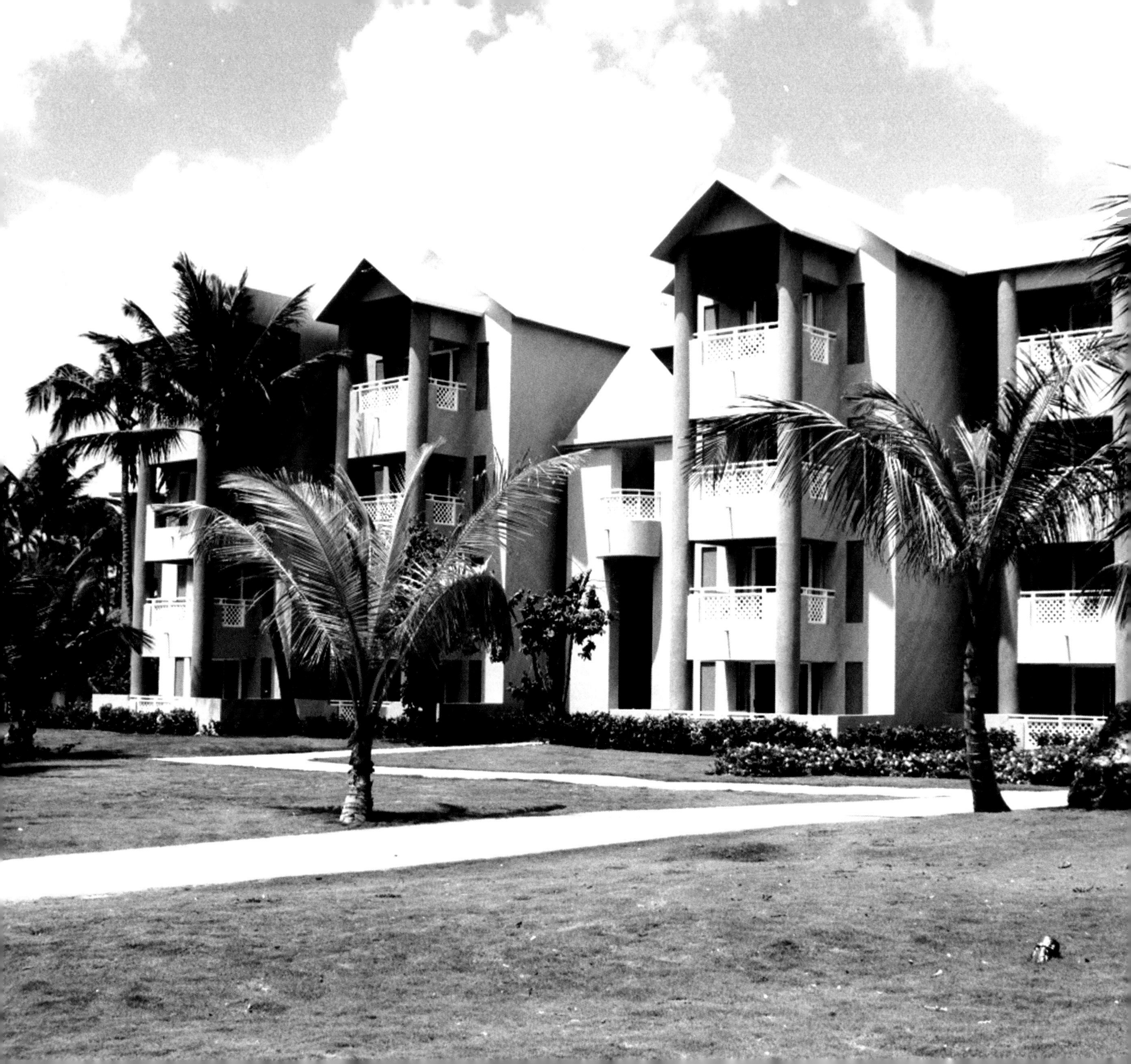

Moving Forward/Blending Vision and Reality

First, there is the vision. Then, comes the task of making the vision real. Then, there is the task of continuing the vision.

In 2005, Punta Cana is still an unfinished experiment. What has been completed is remarkable in itself. An island resort, a cluster of magnificent oceanside homes, a center for the study of biodiversity, and a vibrant, emerging community with increasing modern amenities. Education, jobs, and entrepreneurs.

It is possible to bring about economic development by using a community's indigenous resources. And the outcome can be more widespread and lasting than programs that have their origins elsewhere. But such development requires time, passion, and commitment. Trying to speed up the process can only create half-measures.

What is on the drawing board? Can the succeeding generations continue the unique vision?

In Punta Cana, the initial vision came from two people, unlikely partners. One, a New York lawyer involved in mediation and labor. The other, a Dominican, with an insatiable desire to create something new and lasting. Their first idea—some cabanas that were almost unapproachable—didn't last too long. But as the area changed—the two helped to finance and build a modern airport—the earlier idea became more viable.

With time, that vision changed and evolved too. In the 1990s, the Earth Summit at Rio de Janeiro triggered the need to become more conscious of the environment and the need to protect it. At Punta Cana, that became the basis of an ecological foundation and a center for the study of biodiversity. As the political climate stabilized, so did the currency, increasing the attractiveness of Punta Cana as a resort and location for a second home, in the end, attracting a whole new set of clients and investors. Then came the development of the sur-

and the achievements of Dominicans. Within weeks the plans were drawn. The new center is expected to be ready by the fall of 2005.

Punta Cana is first and foremost a business. But it is also a complete community that has created a wealth of opportunities for those who work and live there. And it plans to go a lot further. Not only do Kheel and Rainieri see it as an even larger and more established community, they are convinced that the formula can work in other countries as well.

In spite of its growth and its continuing success, Grupo Punta Cana is still a family business. The presence of Oscar de la Renta and Julio Iglesias has helped expand the investor base and the oversight, but it will be up to the Rainieris and the Kheels—the controlling shareholders—to decide how to expand and in what direction.

At Punta Cana, Ted Kheel and Frank Rainieri are still the visionaries in charge and their families share their passion and their goals. But as the company grows—both within the Dominican Republic and abroad—the demands on the company and the management will clearly change. How dynamically the new generations respond will determine the continuity of the vision and its evolution.

And what about social and environmental responsibility—the cornerstones of Punta Cana's vitality? Can big businesses reconcile their needs with the needs of community building and environmental stewardship? Does the need for financial returns often push social and environmental concerns to the back? Will Punta Cana be different?

rounding community—new homes, schools, shopping areas, new stores.

How do you create permanence beyond being just a resort? At Punta Cana, as with many other projects of its kind, the future rests a great deal on attracting

the second and third home buyers—retirees looking for vacation homes or residences where they can spend some of their time—and the global resident, one who wants the flexibility of having footprints on several continents. One key element of that has been to create world-class golf courses and luxury home sites overlooking the courses.

There is even more. After a *New York Times* article bemoaned the paucity of "culture" in Punta Cana, Kheel and Rainieri quickly set about correcting that with plans to create a cultural interchange that would showcase art and culture

MOVING FORWARD/
BLENDING VISION
AND REALITY

Frank Rainieri

WE HAVE A LOT OF THINGS GOING AHEAD. Where are we going to get the money? Interest rates are so impossible. One way that I'm very lucky is that my partners, all of them, have agreed that we will reinvest every penny. We don't pull our profits, we reinvest everything. There are very few companies where after 20 years people reinvest everything. Normally, the thinking is: Let's get the money out. But I think that everyone is so proud of what has been accomplished they are happy with seeing the results more than anything else.

I believe that Punta Cana should be multiplied by 10, 12, 15 Punta Canas around the Caribbean. And South America. We know how, we have been able to establish a plan. Here is something that a big packing company in Ohio has been able to develop. How many companies have done so many things at the same time? So we have the know-how that we should export. I think we can help many different areas of the country and of the Caribbean. Raise the standard of living—the highest per capita income in the DR is in this area. We have eight million people. For a Caribbean country, we have the largest population after Cuba, so we are a big country. We have the largest per capita income of the DR in Higuey. It's the only part of the country with no unemployment.

If we could replicate this in five, six, seven different places, do you know what that would mean for our countries?

How could this happen? First, I have to finish setting up my team. As I said, we always did everything on a shoestring. Now is the first time we have a team doing things with 21st-century management. So it means that now we are ready. We have to be.

As soon as my partners—Julio, Ted, and Oscar—agree, we start the second stage. We should go to other places and do the same thing and become a multinational. Of course we don't make the mistakes we made at the beginning because we learned, sometimes the hard way. I think it would be very important to spread this gospel—having other Punta Cane's doing what we do here for so many people around the Caribbean.

Can such things scale up? What are the challenges? I believe the most important thing in life is to know what your target is—don't get distracted by others, what they do, or how they do it. Look at them. Hear them. But, think what your mind tells you to do to reach your objectives.

How big can we grow at Punta Cana? As big as we want to. We still have 5,000 acres—that's a lifetime job. Probably none of the partners will be alive at the time that is totally done. Second, I believe we can reproduce this in other places. Maybe not as big as what we had originally, over here, but at least over 2,000 to 3,000 acres.

Would we like to go public? I never say no to anything because who knows what will happen in the future? Experience has not been good with companies of the Caribbean and of the tourist sector. Services like tourism do not provide high returns for a company. Club Med was a fiasco. I can mention a dozen companies that went public and they went nowhere.

Being a company that has the purpose of sustainable development is a handicap in the market and being small is a handicap. I think that what we should do in the future is that we will have a holding company. Let's say that we go to another island in the Caribbean. We should not be the only investors. We should have preferred shares that will give us the management and we will come with 30% of the investment. Those who don't have the know-how will come up with 70% of the investment.

That would give us a lot of return and a lot of value in our holding company but not require a big capital investment. That's what I believe should be our next stage. We can show books that would give comfort to anybody who wants to invest anywhere. If we go, let's say, to Cuba. Let's say Cuba was free. And we bring thirty to forty investors who each come up with one million dollars. We're not going to show you a slide presentation of who we are. We're going to take you to Punta Cana for a weekend to show you how we operate, who we are, and then at the end we'll show you our results. I think that a lot of people will say that we can deliver because they have seen it. We won't be selling a dream. We can show that our dream has been accomplished.

So that's the way I think we should grow and I believe we have to grow. I think it would be a tremendous mistake to stay where we are. I think Punta Cana cannot be seen as a short-term investment. This is not the kind of company that is for speculation or for short-term investment. Why? Because the basis of this company is two things—real estate and tourism. In 20 years, two major countries of the world—India and China—almost 50% of the population of the world will be more developed than they are today. Today, only one million or two million people in those countries are involved in tourism. Eventually, 30% of the people in those countries will start to do tourism. That's 600 million more people.

Communications are better every day and cheaper. It costs less today to take a charter from New York to Punta Cana than it cost when my father sent me to the U.S. in 1960, 44 years ago. People every day are looking for more and more vacation homes. There's no more land being grown in the world today. Real estate is going to become more and more costly and more and more lucrative. There's only one Caribbean. So the two key elements of our product—real estate and tourism—the potential market is there and that's why I believe that we should start a second Punta Cana in the near future, and a third. Punta Cana should be a style of development—an integrated style of development—not just simply one place.

We're not talking about a resort that is just the hotel. We're talking about how you get there, how you get to the airport, security in the area, what services are provided around the hotel. What is the impact on the environment?

I know 60 major airlines that fly here. I know all the major tour operators of Europe and the U.S. I have access to them. I didn't know any of them before. The know-how is not only how to build a runway or a hotel. We have everything. I can get into my computer and call a dozen major tour operators and I can tell them: "Listen, I'm building a 300 room hotel and I want to start the next place." I can talk to a dozen chains and see which one of them wants to build the second or third hotel or I'll give them the land for this. The same way I've done it here.

If you come here today to buy an acre on the golf course, you pay more than what Club Med paid Ted and me for 20% of the shares of that hotel plus the land plus all my work. That lets people know that they better get in at the beginning because prices will go up.

I'm very happy and proud because normally children don't like to come and work in what the parents have done. In our case, it's Mom and Dad—between my wife and me we have been working in this company for over 50 years. Normally, the tendency is for children to reject that. The other consideration is being isolated from their natural environment—the capital cities where they have all their friends—all the discos, bars, everything.

To come and live in Punta Cana is a big sacrifice. Having a good education—U.S. college grads—master's some of them—to go and live in Punta Cana means that from the day they were born they were filled with the love of Punta Cana that we have. The most important part is to know that there are people in my family who love Punta Cana as much as my wife and I do.

Here the message is: Ted is 91 years old, Julio is 63, Oscar, 73. I'm 59 but there are other people, shareholders coming along who are only 30 and 28 and 26 who have the same view, same love and are willing to continue if they are allowed to and who have the capacity. They are making big changes. Their mentality is not mine. They are more open. They make more changes than I could. They are less conservative—naturally you become more conservative with age. They are more aggressive and I think that has been very important. Things are being done better and faster because of them.

Francesca Rainieri

WHEN WE WERE DECIDING what our new brand would be they asked me: What is Punta Cana to you. I told the person, "Punta Cana is Punta Cana. It's part of our lives. I say our lives because I know it's my whole family. When you grow up somewhere it grows with you, it's in your veins, in your blood, it's not like a normal business." I told my father afterward because we had been thinking about it and I wanted it to be really clear. We've been working as a family. For me it's hard because I can't separate one thing from the other.

I remember when we were little, Dad would take us for horseback rides over all the land—there wasn't anything here. It would take six hours on horseback and we would go through the whole property twice a year, every year because he said we had to look at the land and there was nothing there but just the land—jungle—and it was his way of making us love what he loved as much. So when you see something and you are part of it, it's just part of you. Personally, I always knew that I was going to work here. It's something I am proud of and I am looking forward to it.

Paola Rainieri

I AM THE ELDEST so I am the one who remembers most of how Punta Cana used to be. I remember perfectly the little airstrip that we landed on before there was actually a real airport. I remember the little jeeps that used to take us from there to the hotel. I remember the original cottages. We had a turtle. It's now prohibited to capture them or kill them but at that time we had one at the hotel, a pet, and a little pool that we made for it. The walkways were done with conch shells.

Sometimes you get so involved in everyday work that you tend to forget not where you started, but how it used to be, how simple it used to be. Not long ago, I went for the first time to the place where the new golf course will be, which is actually where the old airstrip used to be. I hadn't been there for over twenty years. I was saying how it felt so good to remember how it actually was and to see what it is today. I felt really proud of what my parents and what we as a family have accomplished. We have grown through this. Maybe we, the siblings, haven't worked here forever, but we've been here and we've helped ever since we've been alive. We've been part of it somehow. It was just amazing to see the natural beauty of Punta Cana, but at the same time to remember that it really was nothing. It was just really dry land.

Francesca Rainieri

AT THE END OF THE DAY we all knew we would come back to Punta Cana, not because we had to but because we wanted to. When the hotel opened, we would all help, sell in the stores. I worked in the kitchen, helped in the kitchen, audited the finances of the bars. We've always been involved.

Paola Rainieri

I wanted to study hotel management but my father sat me down and said you study whatever you choose, I've always told you that. But if you ask me, don't go for a major that restricts you so much. You will always feel like today, tomorrow and the day after, you will always have to work in hotel management. I don't want you to feel that way or to think that you have to work in this area. I think that you should study something that is broader so that tomorrow, if you get tired of hotels you can do some other thing. I was shocked that he told me that because I always thought he wanted me to study hotel management. But when I thought about it he was right so I went for something he didn't expect, which was communications. It's definitely broader. I can practice it here. I can practice it anywhere, within a company or independently. I think he just wanted us to do what we wanted whether it was here or anywhere else so we would never be restricted to working here if we didn't want to.

I studied finance and entrepreneurship. It was too narrow. Finance. I figured if it doesn't help me with what I do in the world, it will help me in my personal life. When I arrived in the company I asked: Where do you need me?

My father never insisted that we had to come to work at the company. Actually, he made us feel that we had to do things on our own. We all had to spend at least two years working somewhere else. They all made us feel since we were very young that we had to build our own life, make our own life. It's a little bit of a contradiction. They were always telling us make our own life but at the end of the day we have the business. And because there is a lot of space for entrepreneurship, we've been able to make our own space within the company.

Francesca Rainieri

When I came to the company the first year it was because Dad said, I need you; it's time to come. I moved my family. I went to human resources. I didn't ask how much I was going to get because you're not doing this for money. Then my paycheck came. I looked at the check. I went to the HR guy and said, "What is this?" He said, "It's your pay. I said, "I'm getting 20% less than I got in my old job." At the end of the day we do this for love. None of the employees work as many hours as we do—yet we probably make less than those in the same range.

My parents are very conscious about never wanting any sort of misunderstanding among the partners. They think that we need to make our way in order to appreciate it. A lot of my friends got married and the parents gave them money for a house. But they won't give us anything unless we work for it so that we respect the money. That's why, for our generation, Punta Cana is safe in that way. Because we remember everything and we lived through it. But we have our kids and the big challenge is passing on the values to the third generation.

My son goes to the school here. He's the grandson of the owner. I tell the teacher to treat him like anyone else. But at the end of the day, even the kids are aware of it. Still, we are trying to help them stay real. We are lucky because they are in this kind of community. They are not as exposed as other kids to as much. Here you are part of the flora and fauna—you don't live so much in the material world. We live in a community where they go to school with kids from every socio-economic level. They come out and go to a house and they don't think about anything material. In that sense I am luckier than my sister. Her kids go to school in Santo Domingo. There's more competition, social status. But the challenge now is to try to make them feel and know what we came from, what we are, and what's special in their life and in the world.

Paola Rainieri

I won't have to tell my kids anything about the business. On Sundays they like to go to the airport and watch the planes. They are growing up here. They go to the beach and play. My daughter is only five, but she knows when something is new or different. She plays at the hotel every day. She knows the people there. She is part of it as she is growing. By the time they are 13, they are going to be quite aware of what it is and how it works. They can start working as soon as they are mature enough, even as a ball boy or a bell boy. I want them to go through all the areas. When they ask why Mommy works, we talk to them and make them understand our need to work for things. We're hoping they will learn by example. It's a working playground for them.

We're all still growing and I think we'll keep growing over the next decade. Many companies reach a point where they level out. In our case, I don't think we'll see this for a long while because we have so many plans. The next three years you'll come back and say: "Oh my god." The residential part, the resort and the residential areas for people who are moving close to the resort in the areas next to us and are now moving to the town—these will grow tremendously over the next three years. There's so much construction, so many people buying, so much being built.

What we've accomplished up until now is very much, but it's still the basic part. From now on, because of the real-estate development, you'll really notice the community growing. It will look so much more complete.

Francesca Rainieri

The biggest challenges will be the next three years. First, we're doing a family division that we haven't done before. It's hard at the beginning. We've been able to start making a structure in the company which was needed. It couldn't be just "Mom and Dad" anymore. Now it's not just us three and my husband (which is four). It's also executives—real executives. We're starting to build a web and when the web is strong, you can stretch it, and stretch it, and stretch it, and it will go anywhere. We're not going to stop working as a family team. But we are starting to be more of a company, a real company. We're completing the team. It has to be a team. It can't be just Mr. and Mrs. Rainieri making the everyday decisions in everything in the company.

Paola Rainieri

There is a vision that combines growth of the resort and the community and everything is inter-related. We've recently made some changes to Mission/Vision/Values statement because we've grown in so many ways as a company. The fun part is that we've grown and within our mission we've added responsibility—social responsibilities that we didn't include before. In our mission we talk about respecting the environment; now we've added social responsibility. Although we were doing it before we didn't really have the definition for it back then. But the people who are the new executives, including us among them, we value this effort that has been made to help the ones who are close to us. And we actually believe that we had to add that to our mission as a company because it is part of being here, helping others and anyone that comes into this company, no matter the title they come with, or the money they re getting paid. They have to be part of this vision and they have to understand that they are here to do business, but also to help others.

Frank Elias Rainieri

Frank Elias Rainieri or Frank Jr. as he is referred to by many is currently part of the real-estate sales team at Punta Cana. A graduate of Boston University and a MBA, he returned to Punta Cana to become involved in the resort.

PUNTA CANA IS A DREAM, not just a development company—a dream that has taken 35 years to build. Punta Cana is always going to be our legacy. A dream that is going to go above what we will try to do anywhere in the world.

I work with my heart here, not with my money or my business expectations and I try to create a legacy that will stay forever. Ten years, twenty years, one hundred years from now, anyone who has a house here will say, "Wow! What a project!" A hundred years from now, people will tell our children and our children's children, "Your great-grandparents did something that no one has done here ever."

Everybody knows everybody here and you have access to everyone. Anybody can go to my office just knock on my office and say: "Frank, this happened to me. I need help from you." The best ideas always come form the bottom, from the lowest employees—the greeters, the waiters, the captains—because they know the day-to-day issues of the business. The good ideas always come from the middle and lower management. We can have the vision of where we want to go, but if we don't have the ideas from the people at the low end we don't have any development.

Another thing that is very important for us going forward is the educational system. We have a great school in order to sustain the development of the area. Without education and without all of our employees speaking a second language like English or French, we are going to be finished. Because in today's world people want to go to a place where they can be understood right away and where they can feel and say what they think. They don't want to go to a place were no one speaks English or German or whatever. With the school that we are creating, we are breaking the education barrier for our employees.

We are starting a program now for housing for all employees. Most of them—80%—live in Higuey, which is an hour away. We are trying to create a big residential area for them. We are going to subsidize the housing, but it will immediately attach them to us and to the company. They won't come here for two years or three years. They're going to be attached to the company for 10 years. We will have better service. Because when homeowners come for a second time or a third time, they already know the people, because they have been here for many years and they feel that they are a part of it.

They'll give better service if they can live here. At Higuey, they don't have electricity, they don't have water, they don't have a nice house. When it's raining, they get wet. Here, they are going to have water, electricity, and a nice house, where the roof is not going to leak, and have good schools for their children. So they are going to be happy. And when they are happy they transmit that to our client—the people who are buying here and living here. So for all of us, those things are very important.

We plan to take what we have done here and go abroad. It's about development. We are going to create a golf course to the same standards that we have here. We're going to have hotels and houses. But it's not going to be Rainieri, Kheel 100% involved. We are going to be involved, but more as managers. And we won't be there forever. So it will be a different kind of development. It's going to be more to help others develop a rural area become a nice place. It won't be the same development that we have here and, of course, it won't take the same time. It took us 35 years to develop here and it will take 10 years to develop in other places. That's because we have the know-how. And we have the connections to go somewhere else.

One of the important things we need to maintain in order to go forward is the partnership between the Kheels and the Rainieris. No one of the four partners controls the company. In the Rainieri case we are in the second generation. In the Kheel case they're in the third and fourth generations.

There is a great deal of responsibility that we have in our hands. No one of the partners had this. At the young age of 27, I have a greater responsibility than any of them had. When my father was 27 he was starting. But I don't have anything that is starting, I have something that is built, so my responsibility at 27 is far greater than Frank or Ted had. At 27, I am part of a $200 million company. The responsibility that my sisters and I have to continue growing the company is more important than anything. But it's a nice challenge. We know we have to do it. From the moment we got involved, the company has jumped two or three stages. It's a completely different company and the company has changed from Club Med to the hotel, to real estate, and in ten years it will be a service industry. So it goes from change to change and we have done the changes correctly. For my sisters and myself, this is not a business, it will always be a dream. We are building this as a legacy.

Ted Kheel

I'VE GIVEN STOCK IN EQUITY TO MY CHILDREN over the years. My wife has done the same. Last year, we had given away enough so that we were minority stockholders. This year, I completed the process and no longer have any financial interest, but I remain the CEO.

Today Punta Cana is totally owned by my children, grandchildren and two great-grandchildren. They are the equity owners of my share of Punta Cana.

Robert Kheel

I MISSED THE FIRST WINTER AT PUNTA CANA because my number one son was born in November 1971, the first time the family went out there. We did go down in 1972. It was quite an experience. I remember being there several years at Christmastime because that's the only time we were there. I remember landing there in the dirt runway. One winter my mother went by jeep.

It was hard to get TV. I remember the village too, but it was the human part I remember most. We had serious water problems because the sanitation wasn't that good. All of us got dysentery or diarrhea. But Dad didn't want to acknowledge that he got it too. We had nice times. But it was primitive and there wasn't much to do. How hard was it to communicate? We had a radio phone, no telephone, limited water, and not exactly the best food.

Every winter some of Dad's friends would come down from New York City. Then they sold it to Club Med. My biggest memory was that the Club Med's structure was to assign you a table as you went in and so you ended up being next to an

Italian or a Frenchman, which was all very good, except that they didn't speak any English and we didn't speak any Italian or French. Mother beat the system by sneaking in and freezing whole tables. Then in the late '80s they built the current hotel and we started staying there and it evolved from there to what it is right now.

Every year, for years, we used to have a stockholders' meeting. It was a combination family gathering and shareholders meeting. Then when Oscar came in, it became more formal. Then the stockholders meeting became more of a family gathering than stockholders meeting. And the Board of Directors meeting became a different but more important function.

There was a joint venture where Oscar and Julio were partners of a Corales partnership that was really a sub-partnership. That's when they first came into Punta Cana. Then, as part of the negotiation to buy out Club Med's share of the airport, Oscar was of great assistance and as a result of their contribution a decision was made to merge their interest in Corales to a stock interest in Punta Cana. As a result each got 8%. In 2000, we had another negotiation to increase their stake.

Punta Cana has become a remarkable place. But there are still a lot of challenges going forward. There's the business and there's the governance.

Grupo Punta Cana is no longer a mom-and-pop shop and it's not the same business. At one time, Frank and Dad could run the business on a phone call and a prayer. But now, this is a major business.

Frank has been incredible but he needs—and he recognizes it--to build up a level of senior management. Going forward, we need to develop institutional mechanisms and staff. To operate a business as big as this is a major challenge.

The development of the Punta Cana town is a central aspect of the growth because you need the new managers to live in Punta Cana, be committed to the community, and be a senior level of leadership. We need that. We need the families and the board of governors to be more formally involved. It can't just be Dad, Frank, and a phone call.

Human dynamics. You put four people in a room and they won't agree on everything. Where it used to be easy, it's more complex. There's a complex development of what is being a major business—a serious business. It's got great risks and even greater opportunities

It is obviously a lot more complex than before. There's that great phrase. If necessity is the mother of invention, then invention is the mother of necessity. The very success we have had has created incredible demands and pressures and we have to be able to respond to these in terms of our community, the barons, the currency problem, and the malaria problem. We have to be able to respond to all those problems and that requires a thoughtful and committed business strategy.

There's a human institutional component that is very important. That's where Ernie, my brother-in-law, and I are most able to function. Arnie is a securities lawyer who specializes in advising boards and I am a litigator who sees things when they get messed up. We come at it from opposite ends. To Frank's credit again he is very responsive both consciously and intuitively. He recognizes the need to a more institutional approach to things.

The stakes are much higher, the decisions are more dangerous, and the potential benefits are much greater. I don't think we are ever going to abandon a very conservative strategy toward growth. We've never wanted to be highly leveraged. Never consciously wanted to expand beyond what we thought was conservatively prudent at the risk of losing some benefits because we don't want to be in a situation where you have to do something—have to sell or whatever—and I am confident that this is going to continue.

The decisions are more complex. The demands are just staggering right now. When they build this new road to the North Coast, the demands are going to get even greater. What do you do with the airport extension? What do you with the water? What do you do with the electricity? You're talking tens of millions of dollars' decisions. They are very complicated and Frank's been using a lot of experts and advisers to make sure that the answers add up.

The future is a conservative growth path. All of us are not adverse to money but no one's in it the way they would be in IBM stock. We all love the place. Would I rather have a dividend than stock? I'm going to find a mix. I don't want to use this as a cash cow and milk it. I want to use it to continue to grow patiently.

I see some of our family at all times having an oversight role—Arnie more so than I because he is very well connected. We do have involvement in getting people involved in Punta Cana. The real question is the next generation of leadership on the operational level. To me the Kheel family is committed to Punta Cana. We have no interest in exiting.

Epilogue

There is a school of thought in development that argues that economic change—especially in developing countries—cannot happen without wiping the old slate clean and starting afresh. Punta Cana is proof that there is another way.

THE FOUNDERS OF PUNTA CANA started with a basic premise that development, especially one built around tourism, had to be a sustainable process. Sustain the environment and you build tourism. Indeed, the uniqueness of Punta Cana lies in its environment and nature and the community that exists around it. And the inability to maintain this uniqueness would not only hinder development, it could also prove counter productive.

Tourism is like any other business, says Ted Kheel... It has its raw materials – the weather, the water, the sun, and it has its factory—the environment. Environmentalism in tourism is the maintenance of the basic equipment. And sustainable tourism is about how the environment is treated along with the community, the people and the guests.

People come to places such as Punta Cana for the weather, the water, the sun. But it is the environment and the community around it that gives Punta Cana a different dimension. "We are attractive to people because of the respect with which we hold nature," says Kheel. But the attraction is also because of the manner in which the culture of the resort is

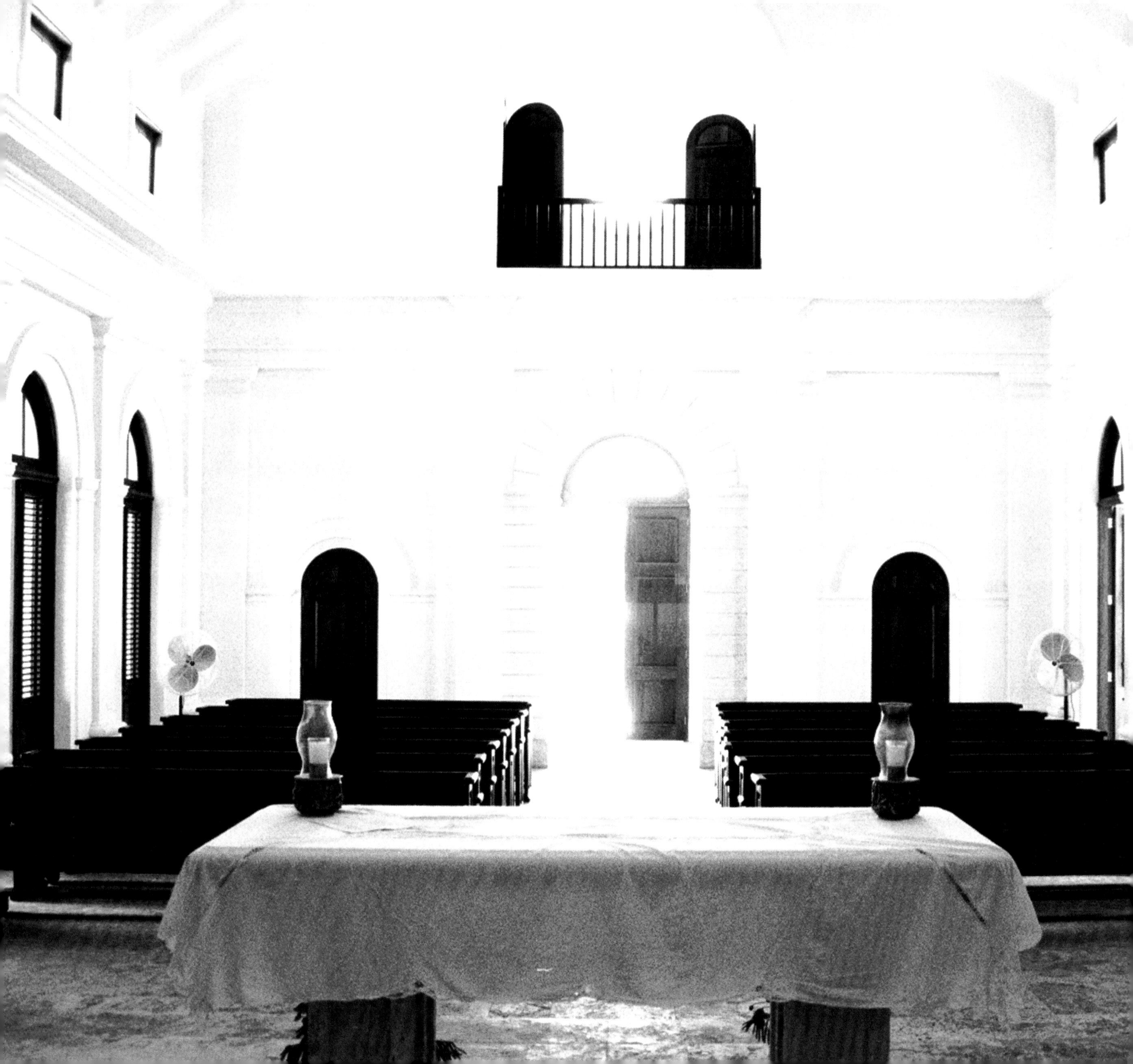

really the culture of the community.

A reporter from the Washington Post in 2004 commented that if you are looking for Dominican culture, don't look for it in Punta Cana. She was right. Punta Cana's culture is not Dominican. But in the years since Punta Cana began, there has been a community, a church, a school for the children of the workers and the residents, a Polytechnic to provide technical education, an ecological preserve and now a cultural center that will be home to a library and an arts facility.

Punta Cana has a culture of its own, a culture that has come from the many influences that have shaped it. And while that culture may not be traditional Dominican it is a culture that is uniquely Punta Cana.

"We are about balanced and sustainable development," says Frank Rainieri. Punta Cana is a resort that cares not only for the hundreds of thousands of visitors that come each year and the hundreds that have chosen to make it their second and third homes but also for the thousands that live and work there. There is a concern for the totality ...a concern that makes Punta Cana and its developers more social scientists and architects than a group that simply wants to reap the benefits of the weather, the water and the sun.

As we emerge from a global economy that has been dominated by a few developed economies into a truly interdependent global economy, the lessons of Punta Cana are simple. Sustainable development is possible, even in the poorest and depressed communities. But the seeds of that change and development have to come from internal resources and assets, not from externally imposed ideas and technologies. And although the nature of development in Punta Cana is not "disruptive" as many economists and development experts believe it should be, its balance and its evenness is a lesson for us all.

Not only has Punta Cana grown from the initial handful of villas and an experiment that brought skepticism from the Wall Street Journal, it has become the catalyst of a development process that now touches dozens of communities that surround it. And the development isn't simply business, it is social, religious and educational. Not an easy task in a country where the largest exports are baseball players, alcohol and tobacco.

Where will Punta Cana be ten years from now, or even five? How will it compare to other developments in the Dominican Republic, in the Caribbean and in other similar island communities? Whatever happens, the impact of the last thirty five years has been unmistakably positive. It can only get better.

AEROPUERTO INTERNACIONAL DE
PUNTA CANA
PUNTA CANA RESORT & CLUB
www.puntacana.com

Punta Cana Time Line

1968

UNESCO declares Punta Cana beaches exceptional and unique in the world for their fine sand and crystalline water.

1969

At midwinter meeting of AFL-CIO in Miami, Ted Kheel hears about inexpensive land for sale in the Dominican Republic. He assembles a group of 40 American labor and management investors who purchase 30 square miles of raw jungle and beachfront with no access roads and no infrastructure.

1970

Kheel meets Dominican entrepreneur Frank Rainieri who becomes a partner and local representative of CODDETREISA, the acronym of the holding company that is formed to develop the property. The name is later changed to Grupo Punta Cana.

1971

Rainieri builds a small resort on the northern part of the property. Punta Cana Club is a complex of 10 small cottages and a Clubhouse. Guests fly in on dirt airstrip. President Balaguer attends the dedication.

First Wall Street Journal article appears and incorrectly reports that American labor leaders are financing a cotton plantation in the Dominican Republic and paying substandard wages. Actually, one man was trying unsuccessfully to grow cotton on a small patch of soil.

1972

Rainieri creates first small school at Punta Cana for workers' children.

1973

Puerto Rican caretaker and American partner sells off property, later transferred to a Dominican, without permission. Kheel & Rainieri begin a three year lawsuit in the Dominican Republic to reclaim the land.

1975

Dominican trial and appellate courts decide in favor of Kheel, Rainieri and CODDETRIESA.

While Punta Cana Club hardly covers expenses, it does demonstrate the potential of the area for tourism. Conversations begin with Club Mediterranee, then searching for a site in the Dominican Republic. Tentative deal requires CODDETRIESA to find financing for Club Med, which gets land for a modest fee.

1977

Club Med moves forward with the deal, paying a mere $300,000 for the original Punta Cana Club land. Club Med begins 3 years of construction on a large tract of land to build a 350-room hotel.

1978

First Punta Cana master plan created by Dominican architect, Oscar Imbert, as master's degree project for Pratt Institute of Technology in NYC.

Rainieri convinces Dominican government to begin work on first paved road to connect Punta Cana with nearest town, Higuey, 35 miles away.

1980

Club Med launches resort but discovers access is virtually impossible without an airport long enough to accommodate overseas flights.. Gets turned down by the Government for its lack of funds. Turns to Kheel and Rainieri to create a joint venture.

1983

After negotiations with the government, authorization obtained for the construction of an international airport permitted to charge passengers landing at the airport.

Rainieri oversees building of the airport. Oscar Imbert designs unique open air terminal with thatched cane roof using local materials such as coconut trunks, local trees and cana leaves.

1984

Airport inaugurated with first international flight from San Juan, Puerto Rico

1986

Groundbreaking on new resort with group of Dominican partners. Rainieri finds funding using recent government tax exemption plan to promote tourism.

1987

Overseas Private Investment Corporation provides financing for airport extension to accommodate long distance international flights. With cash flow, airport extended to 9,200 feet and then to 10,400 feet, opening the door to Trans-Atlantic jumbo jet charters direct to Punta Cana.

1987-1988

New hotel, Punta Cana Resort and Club, opens. Signature structure, La Tortuga Beach Club is the largest thatched roof structure in the Dominican Republic, helping to revive this form of traditional architecture.

1994

Second Wall Street Journal article appears, praising resort as a remarkable development feat. "Punta Cana Turns Jeers into Cheers."

Punta Cana Ecological Foundation formed. This non-for-profit organization manages 2500 acre conservation area, interpretive center and network of nature trails.

1995

Punta Cana hosts first Caribbean Conference on Sustainable Tourism

1997

Oscar de la Renta and Julio Iglesias become partners and build homes in the new Corales estate area. Rainieri and Kheel follow suit.

Grupo Punta Cana purchases Club Med's half of the airport.

1998

Marina completed and plans begin to sell lots and private homes. Hurricane George destroys large part of the area. Resort rebuilds in 3 months.

2001

Opening of the Punta Cana Biodiversity lab which develops collaborative relationships with seven American universities and numerous Dominican institutions and community organizations.

Resort opens P.B Dye's La Cana Golf Course with magnificent beach and sea views.

World Summit on Sustainable Development in Johannesburg, South Africa, selects Grupo Punta Cana as a model of sustainable development.

2002

Grupo Punta Cana begins to contract out services to address the increasing challenges of an expanding business.

Dominican magazine MERCADO selects Grupo Punta Cana as the most admired tourism corporation in the Republic.

2003

The Plaza shopping area is opened, construction begins on worker housing, new elementary school and church are built.

A New York Times article describes the resort as a socially responsible business "with the kind of corporate responsibility more often found in companies in the United States."

2004

First phase of Ann & Ted Kheel Polytechnic (High) School completed for up to 500 students from the surrounding community. Dominican President Lionel Fernandez Reyna attends the dedication. Secretary of education Alejandra Germen expresses hope that it will be a model for other provinces.

2005

With passengers up from 2,000 people in 1984 to a current 1.29 million, Punta Cana International Airport opens a new terminal.

Frank Rainieri receives lifetime achievement award at the Caribbean Hotel Industry Conference in recognition of his commitment to responsible environmental and social development at the Punta Cana Resort.